B

BioMethods
Vol. 1

H. P. Saluz
J. P. Jost

A Laboratory Guide
to Genomic Sequencing

The Direct Sequencing
of Native Uncloned DNA

1987
Birkhäuser
Basel · Boston

Cover drawing by Jean-Pierre Jost

CIP-Kurztitelaufnahme der Deutschen Bibliothek

Saluz, Hanspeter:
A laboratory guide to genomic sequencing :
the direct sequencing of native uncloned DNA /
Hanspeter Saluz ; Jean-Pierre Jost. –
Basel ; Boston : Birkhäuser, 1987.
(BioMethods; Vol. 1)
ISBN 3-7643-1925-9 (Basel)
ISBN 0-8176-1925-9 (Boston)
NE: Jost, Jean-Pierre: GT; HST

Library of Congress Cataloging-in-Publication Data

Saluz, Hanspeter, 1952 –
A laboratory guide to genomic sequencing: the direct sequencing
of native uncloned DNA / Hanspeter Saluz, Jean-Pierre Jost.
p. cm. – – (BioMethods: vol. 1)
Bibliography: p.
Includes index.
ISBN 0-8176-1925-9 (U.S.):
1. Nucleotide sequence – – Laboratory manuals.   I. Jost, Jean-
Pierre, 1937 –   II.   Title.   III.   Series.
[DNLM: 1. DNA – – laboratory manuals. 2. Genetic Technics.
QU 25   S181L]   QP625.N89S25 1987
574.87' 322 – – dc 19

© 1987 Birkhäuser Verlag Basel
Typography and Cover Concept: Albert Gomm
Printed in Germany
ISBN 3-7643-1925-9
ISBN 3-8176-1925-9

*Le chercheur qui insiste sur l'importance
de ce qu'il a trouvé, on le juge indiscret.
S'il n'insiste pas, on dit qu'il n'en a pas saisi
l'importance.*

JEAN ROSTAND
Carnet d'un biologiste

Dr. Hanspeter Saluz
Dr. Jean-Pierre Jost
Friedrich Miescher-Institut
CH-4002 Basel (Switzerland)

Contents

Acknowledgements

We are grateful to a number of colleagues both here in the institute and elsewhere, without whose help this book would have been impossible to write. We are especially indebted to Professor A. Weissbach (Roche Institute of Molecular Biology, Nutley, New Jersey), Professor G. Schuetz and Dr. P. Becker (Deutsches Krebsforschungszentrum, Heidelberg) for reading the manuscript, and for helpful discussions. We would also like to thank the following, Dr. I. M. Feavers (F. M. I., Basel), Dr. J. Jiricny (F. M. I., Basel), Dr. I. J. McEwan (F. M. I., Basel), and Dr. A. Milici (M. D. Anderson Tumor Institute, Houston) for spending long hours in discussion, for reading earlier drafts of the manuscript, and most importantly for checking at the bench that the step-by-step protocols work. We are also grateful to Drs P. Kuenzler (F. M. I., Basel) and W. Wiebauer (Siemens, Munich), without whose help the writing of the theoretical section would not have been possible.

Finally, we would like to thank Dr. R. A. Laskey (CRC Molecular Embryology Group, Cambridge) and Amersham (U. K.) for allowing us to use the data in Figure 14, and lastly many other colleagues in the institute who gave their time freely, especially I. Obergfoell (Photography), M. Vaccaro and Dr. K. Wiebauer.

H. P. Saluz
J. P. Jost

Basel, June 1987

I Introduction

A Safety Considerations

Genomic sequencing involves a number of hazardous steps, such as high current, high voltage, radioactive and highly toxic chemicals. It is, therefore, absolutely essential that the instructions of equipment manufacturers be followed and that particular attention is paid to the local and federal safety regulations.

B   Introduction

During the cloning of genomic DNA many of its characteristics are permanently lost. It was therefore necessary to develop a new technique that would give us a closer look at a gene in its normal environment. The powerful technique of genomic sequencing, first described by Church and Gilbert (1984) now makes it possible to have a precise view of a given DNA sequence in a chromosome. This method combines the chemical DNA-sequencing procedure of Maxam and Gilbert (1980) with the detection of DNA sequences by electroblotting and indirect end-labeling by hybridization. Besides studies on the methylation state of single bases in a given gene (Nick et al., 1986; Saluz and Jost, 1986; Saluz et al., 1986), genomic sequencing can also be used to study specific DNA-protein interactions *in vivo* (Church et al., 1985; Giniger et al., 1985; Becker et al., 1986; Ephrussi et al., 1985; Martin et al., 1986; Nick et al., 1986; Zinn and Maniatis, 1986). Protein bound to the DNA protects specific bases against chemical modifications (example: dimethylsulfate) or enzymatic degradation (examples: deoxyribonuclease I; S1 nuclease) resulting in a characteristic gap in the sequence. Finally allelic polymorphism and point mutations can be directly detected without having to isolate and clone all the alleles of interest. This book contains a detailed step-by-step protocol of genomic sequencing, as optimized in our laboratory (Saluz and Jost, 1986). A trouble-shooting guide at the end of the book should help the reader to avoid the numerous pitfalls we encountered in the early stages of this work. By making our protocol available to the scientific community at large we hope to encourage other workers to adapt this very powerful technique to their needs, and thus make it even more widely applicable.

C The Principles of Genomic Sequencing

This technique consists of digesting to completion totally intact, purified genomic DNA with a suitable restriction enzyme. After purification of the reaction products by repeated phenol/chloroform extractions and ethanol precipitation, the restricted DNA is dissolved and subjected to specific chemical reactions for the individual bases according to Maxam and Gilbert (1980) and Rubin and Schmid (1980). The conditions are chosen as to give a partial reaction of about one cleavage in the DNA per 500–700 bases. The products of the chemical reactions are then separated on a sequencing gel (Stanley and Vassilenko, 1978; Maxam and Gilbert, 1980). The DNA fragments are transferred from the gel to a nylon membrane by electroblotting and are fixed to the membrane by a combination of UV irradiation and heating under vacuum. The sequence of interest is visualized by hybridizing a labeled single-stranded DNA or RNA probe to one end of the appropriate restriction fragment as depicted in Fig. 1. For the indirect end-labeling it is of paramount importance that a unique sequence is chosen. Furthermore, the amount of hybridizable internal reaction products (Fig. 1, wavy lines) should be kept to a minimum, as these will impair the resolution of the sequence.

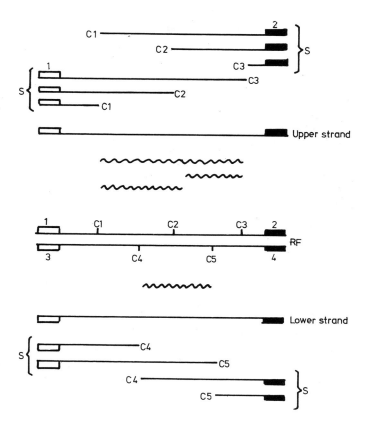

Fig. 1. Indirect end-labeling

For a given restriction fragment (RF) there are 4 possible probes
(in the figure, numbered 1–4). The diagram shows an example
of the product of the C-specific sequencing reaction. By using
the probes 1–4 the fragments designated by S will hybridize.
The DNA fragments created by more than one cut and represented by the wavy lanes will not hybridize with the labeled
probe (1 and 3 open boxes, 2 and 4 filled boxes).

II Theoretical Background

A Basic Theory of Genomic Sequencing

One of the main differences between the classical Maxam & Gilbert DNA sequencing procedure (Maxam & Gilbert, 1980) and genomic sequencing is the way the DNA fragments are labeled. In the Maxam & Gilbert sequencing method cloned double-stranded DNA fragments are directly labeled at one of either the 3'- or 5'-ends. Following the chemical reactions and separation of the DNA fragments on a sequencing gel, only the fragments carrying the label are detected by autoradiography. Thus any subfragments that are not end-labeled do not contribute to the signal.

In genomic sequencing the whole uncloned genomic DNA is completely digested with a restriction endonuclease. All DNA fragments undergo chemical cleavage, separation on a sequencing gel, transfer and binding to a nylon membrane. A radioactive, single-stranded probe is then hybridized to one end of the restriction fragment. This indirect end-labeling enables the sequence of interest to be detected from the pool of genomic DNA fragments. The disadvantage of this indirect end-labeling is that subfragments not contributing to the sequence ladder may also hybridize to the probe and thus contribute to the background. In the indirect end-labeling procedure the background signal is dependent on the following parameters:

P_n: length of the probe in nucleotides
G_n: length of the genomic target DNA fragment in nucleotides
σ: Sigma, the number of cuts per target DNA fragment G_n
E_n: Size exclusion of the gel (in nucleotides). Only fragments smaller or equal to E_n are resolved by the gel.

We would like to demonstrate by means of examples, the influence of the above parameters on the hybridization background. First let us consider the case, where $\sigma = 1$.

Example 1: Ideal Case Condition: $\sigma = 1$

In the case of only one cut per DNA fragment, no background is generated as long as $P_n \leq G_n - E_n$

The maximal allowable length of the probe (for sigma = 1) is therefore:

(I) $P_{nmax} = G_n - E_n$

When $P_n > G_n - E_n$ the fragments b (see Fig. 1) that enter the gel and hybridize with the probe produce a second superimposed sequencing ladder.

In reality, a situation where all DNA fragments are cut only once does not arise. Thus when $\sigma = 1$ represents an average number of cuts per fragment a number of fragments will remain uncut, while others will be cut twice, a smaller number three times, etc. For the purpose of this discussion only integral values of σ will be considered. The following questions now arise: what is the upper acceptable limit of P_n for $\sigma > 1$, and how does this affect the hybridization background? One approach to the problem is to estimate the probability of the occurrence of DNA fragments that will contribute to the hybridization background.

Let us now consider the case representative of most genomic sequencing experiments, where $P_n < E_n$; $G_n >> P_n + E_n$. (A given cut x will be represented by s_x and the subfragment n by S_{nx}, i. e. $S_{nx} = S_x - S_{x-1}$.)

Example 2: $\sigma = 2$

When $\sigma = 2$, four different cases can arise, of which only case a will contribute to the hybridization background.

a)
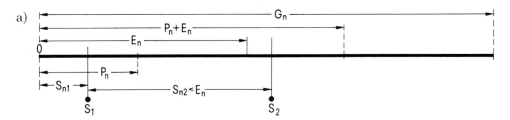

In the case of 2 cuts per $P_n + E_n$, two fragments will be resolved in the gel: one specific and one background fragment, respectively.

b)

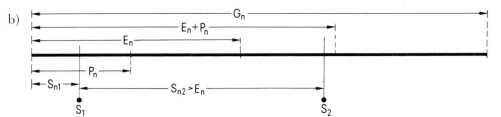

In this case only the specific fragment will be resolved in the gel.

c)

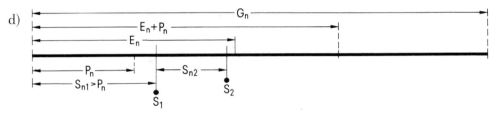

In this case no fragment will be resolved in the gel.

d)

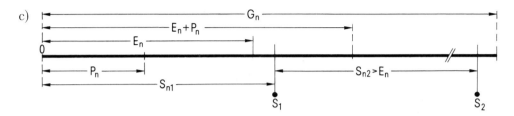

In this case no background fragments are created.

Example 3: $\sigma = 3$

The case for $\sigma = 3$ can be described similarly as for $\sigma = 2$; only one situation giving rise to an increased backround will be dealt with here.

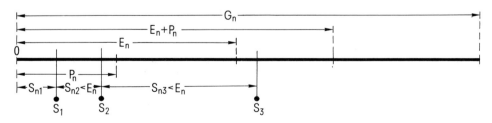

If $S_2 - S_1 < E_n$; $S_3 - S_2 < E_n$ and $S_2 < P_n$, then 2 background fragments and 1 specific fragment will be created.

Based on the above and similar examples two physicists (Drs Wolfgang Wiebauer, Siemens Munich, and Peter Kuenzler, FMI Basel), determined the terms contributing to the background. As a simplification, the parameter σ (sigma) was not regarded as a probability distribution (in addition we have disregarded the fact that there is a minimal length of a DNA fragment required to hybridize). As shown in example 1, where $\sigma = 1$, no 'background' fragments occurred, therefore the derivation given below is made for $\sigma > 1$. The equations for $\sigma = 2$ and $\sigma = 3$ will be the basis for establishing a general equation for any given value of sigma. We assume that $S_1 < S_2 \ldots < S_\sigma$.

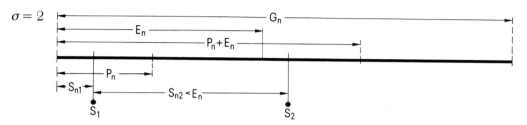

It can be shown (see p. 21) that the probability p $(S_{n1} > P_n)$ can be simplified to

$$p\,(S_{n1} \geqslant P_n) \simeq \left(1 - \frac{P_n}{G_n}\right)^2 + 2\,\frac{P_n}{G_n}\left(1 - \frac{E_n + P_n}{G_n}\right)$$

A background fragment is created only under the condition that $S_{n1} < P_n$ and $S_2 - S_1 < E_n$; otherwise no background fragment occurs (see example 2).

Therefore:

$$P\,(\sigma = 2) < p\,(S_{n1} < P_n) = 1 - p\,(S_{n1} \geqslant P_n) = 1 - \left(1 - \frac{P_n}{G_n}\right)^2 - 2\,\frac{P_n}{G_n}\left(1 - \frac{E_n + P_n}{G_n}\right)$$

$\text{\footnotesize background}$

We define now the probability of background β as:

$$\beta = \frac{p\,(occurrence\ of\ background\ fragment)}{p\,(occurrence\ of\ any\ fragment) = 1\ for\ a\ given\ \sigma \neq 0}$$

then:

$$\beta_{\sigma=2} \simeq 1 - \left(1 - \frac{P_n}{G_n}\right)^2 - 2\,\frac{P_n}{G_n}\left(1 - \frac{E_n + P_n}{G_n}\right)$$

THEORETICAL BACKGROUND

$$\sigma = 3$$

The situation for $\sigma = 3$ can be described in a way similar to that for $\sigma = 2$ (see p. 18)

$$p\,(S_{n1} \geqslant P_n) \simeq \left(1 - \frac{P_n}{G_n}\right)^3 + 3\,\frac{P_n}{G_n}\left(1 - \frac{E_n + P_n}{G_n}\right)^2$$

hence

$$p\,(S_{n1} < P_n) = 1 - p\,(S_{n1} \geqslant P_n)$$

hence

$$p\,(\sigma = 3) < p\,(S_{n1} < P_n) = 1 - p\,(S_{n1} > P_n) = 1 - \left(1 - \frac{P_n}{G_n}\right)^3 - 3\,\frac{P_n}{G_n}\left(1 - \frac{E_n + P_n}{G_n}\right)^2$$

From the above equations for $\sigma = 2$ and $\sigma = 3$ we can formulate a general equation for *small* values of σ.

(II)

$$\beta\,(\sigma) \simeq 1 - \left(1 - \frac{P_n}{G_n}\right)^\sigma - K\,\frac{P_n}{G_n}\,; \sigma \text{ kept small}$$

$$\text{where } K \stackrel{\circ}{=} \sigma\left(1 - \frac{E_n + P_n}{G_n}\right)^{\sigma - 1}$$

- If $P_n \ll G_n$, i. e. a probe that is only a few base pairs long, the dependence of β on the number of cuts (σ) is dominated by K. For practical purposes K varies by less than a factor of two when σ is less than five. This is not unexpected, since a small probe will hybridize with fewer of the subfragments that give rise to the background.
- In practice P_n/G_n will not be negligible (usually $P_n/G_n \sim 0.15$) and so the

    dependence of β on σ determined by $\left(1 - \frac{P_n}{G_n}\right)^\sigma$ is increased drastically.

Although simplified, formula II nonetheless points to the most important factors that influence the outcome of a genomic sequencing experiment.

Thus, successful results can be obtained, provided the following:

1. σ is kept *small*, since the background as estimated by β increases drastically with the number of chemically induced cuts.

2. The length of the probe is not too long because the probability of background signals increases with P_n. Theoretically, 1 residue (i. e. $P_n = 0$) does not increase the background ($\beta = 0$). In practice, however, a probe of a minimal length is needed to obtain a radioactive signal high enough to be detected by autoradiography. This means that the choice for P_n will be a compromise between background and signal strength. To circumvent this problem (which we describe later), we chose a probe of intermediate length with an additional, non-complementary radioactive 'tail' from M13. This increases the signal strength without contributing to the background.

3. E_n, as determined by the acrylamide concentration and the ratio of acrylamide : bisacrylamide, is chosen so that the maximal resolution of a specific sequencing ladder is required. As shown in formula II, an increase in E_n also increases the background.

4. In order to resolve the sequence of interest, the size of the restriction fragment, G_n, must be greater than the sum of the length of the probe and the size exclusion of the gel, $P_n + E_n$.

Mathematical derivation of the generation of 'background' fragments:

$$\left(\text{Definition:} \binom{n}{k} = \frac{n(n-1)(n-2)\ldots(n-k+1)}{1 \cdot 2 \cdot 3 \ldots \cdot k} \; ; \binom{n}{0} = 1 \right)$$

The probability that no background fragment is created:

$$p(G_n, E_n, P_n, \sigma) := p$$

$$\text{(III)} \qquad p = \dfrac{\dbinom{G_n-P_n}{\sigma}}{\dbinom{G_n-1}{\sigma}} + \dfrac{P_n-1}{G_n-1} \cdot \dfrac{\dbinom{G_n-(P_n+E_n)}{\sigma-1}}{\dbinom{G_n-2}{\sigma-1}} \cdot \sigma + \text{additional terms}$$

| Probability that all cuts lie within $[P_n, G_n]$, i.e.: $S_1 \ldots S_\sigma$ $\geq P_n$ | Probability that 1 cut lies within $[1, P_n-1]$, i.e.: one $sx < P_n$ | Probability that all other $(\sigma-1)$ cuts lie within $[P_n+E_n, G_n]$ | Number of possibilities to choose 1 cut of σ cuts | Probability that 2 or more cuts lie within $[1, P_n-1]$ with $S_{x+1} - S_x > E_n$, if $P_n > E_n$ and other possibilities occur (ex. $S_1 < P_{n-1}$, S_2 within $[P_n, P_n+E_n]$ but $S_2 - S_1 > E_n$ etc. |

and the general formula for β:

$$\text{(IV)}$$
$$\beta = 1-p = 1 - \dfrac{\dbinom{G_n-P_n}{\sigma}}{\dbinom{G_n-1}{\sigma}} - \dfrac{P_n-1}{G_n-1} \cdot \dfrac{\dbinom{G_n-(P_n+E_n)}{\sigma-1}}{\dbinom{G_n-2}{\sigma-1}} \cdot \sigma - \text{additional terms}$$

ex.

$G_n = 1000$
$E_n = 300$
$P_n = 100$
$\sigma = 3$

$$\beta = 1 - \frac{900 \cdot 899 \cdot 898}{999 \cdot 998 \cdot 997} - \frac{99}{999} \cdot \frac{600 \cdot 599}{998 \cdot 997} \cdot 3 = 16{,}2\,\%$$

To derive the simplified formula II (p. 19) we assume:

$G_n \simeq G_n-1 \simeq G_n-2 \ldots \simeq G_n-\sigma$
$P_n \simeq P_n-1$

B   Flow Diagram

Total Genomic DNA	The Sequence Standard (Cloned DNA)	Probe for Indirect End-Labeling (Single-Stranded DNA or RNA Probes)

1. Isolation of Genomic DNA
   ▼
2. Restriction Digest of Genomic DNA   Restriction Digest of Cloned DNA   10. Cloning of DNA Probe in M13

   ▼

3. Chemical Sequencing Reactions on Restricted DNA   11. Large-Scale Preparation of Cloned DNA in M1
   ▼
4. Separation of Reaction Products on a Sequencing Gel   12. Synthesis of Oligo-nucleotide Primers and Labeled Single Stranded Probes
   ▼
5. Electrotransfer to Nylon Membranes
   ▼
6. Immobilization of DNA on a Nylon Membrane   13. Purification of Labeled Single-Stranded Probes
   ▼
7. Prehybridization and Hybridization of Immobilized DNA with Labeled Single-Stranded DNA Probes ◀───
   ▼
8. Processing of the Hybridized Filters
   ▼
9. Autoradiography and Photography

III Experimental

1   Isolation of Genomic DNA

Preparation of Nuclei and DNA

Within the scope of this laboratory guide it is impossible to give a full review of all the procedures used in the isolation of intact genomic DNA. However, we shall describe a general method we used which gives an excellent quality of both nuclei and DNA suitable for genomic sequencing. Whether we used cells in tissue culture or specific tissues we found that the best quality of DNA was always obtained from isolated nuclei. If the nuclei are to be used for *in vitro* footprinting it is important to bear in mind that the salt concentration used during the isolation of nuclei can greatly influence the structure of the chromatin (Lohr, 1986; Walker and Sikorska, 1986) and hence the result of the footprint. In this context we also found that traces of certain detergents such as Triton X-100 can greatly affect the quality of the nuclei. For example the presence of 0.05 % Triton X-100 during the preparation of chicken liver nuclei drastically alters the morphology of the nuclei (Fig. 2). Such nuclei have an altered chromatin structure and no longer synthesize RNA *in vitro*. Some cells and tissues contain large amounts of deoxyribonuclease activity; in such cases Mg^{++} ions should be replaced by low concentrations of poly-amines. In this chapter we shall describe the isolation of clean nuclei from chicken liver and describe a general procedure for the isolation of intact DNA.

▶

Fig. 2. Isolation of the nuclei

Effect of low concentration of Triton X-100 on the morphology of chicken liver nuclei. Upper panel shows an electron micrograph of nuclei isolated in the absence and lower panel in the presence of 0.05 % Triton X-100.

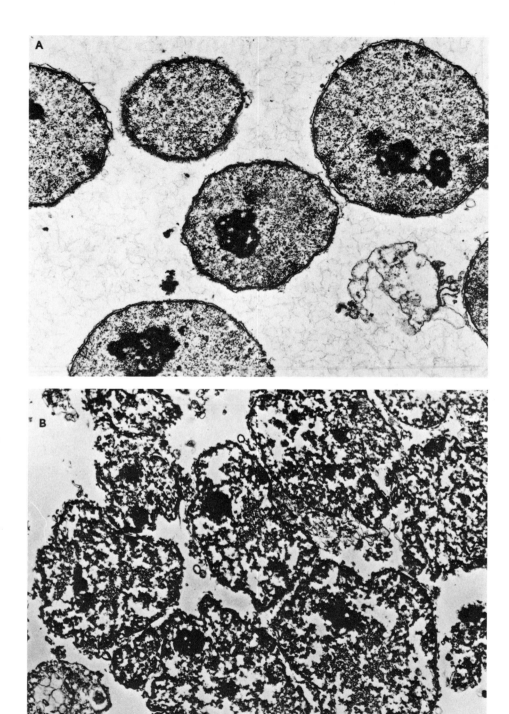

Materials and Buffers

Preparation of Nuclei and DNA

> A loose-fitting glass Teflon homogenizer

> A tight-fitting glass-glass Dounce homogenizer

> A SW 27 Beckman rotor & ultracentrifuge (or equivalent)

> A HB-4 Sorvall rotor & centrifuge (or equivalent)

> Dialysis tubing: the tubing is treated as follows: boiled for 30 minutes in 4% $NaHCO_3$, then 30 minutes in 5 mM EDTA and finally boiled for 10 minutes in distilled water. The tubing can be autoclaved in 10 mM Tris, pH 8; 1 mM EDTA and stored at 4° C.

> Phenol saturated with 1 M Tris, pH 8; 0.1% hydroxyquinoline

> Chloroform

> 10 × Dialysis buffer: 100 mM Tris, pH 8; 10 mM EDTA

> Buffers for homogenization:

    – 80 mM NaCl; 1 mM EDTA; 20 mM HEPES, pH 7.5 containing freshly added 2 mM dithiotreitol; 0.5 mM spermidine; 0.15 mM spermine.

    – 2.2 M sucrose in the above buffer

    – 1.8 M sucrose in the above buffer

    – 0.35 M sucrose in the above buffer

> Nuclei storage buffer:

    – 50 mM HEPES, pH 7.5; 25% glycerol; 2.6% bovine serum albumin; 0.1 mM EDTA; 0.15 mM spermine; 0.5 mM spermidine; 2 mM dithiothreitol

> DNA preparation buffer:

  – 20 mM Tris, pH 8; 20 mM EDTA; 1 % sodium dodecylsulfate (SDS)
  – Proteinase K 50 mg/ml
  – Pancreatic ribonuclease: 10 mg/ml in water. Contrary to common belief, heat treatment of the ribonuclease at 80° C for 30 minutes does not fully destroy the contaminating deoxyribonuclease I, hence such heat treatment is unnecessary and can be omitted. The best way to selectively inhibit any DNAse I contaminating a ribonuclease A preparation is to add 10^{-3}M EDTA and 10^{-4}M EGTA to the incubation mixture (A. Weissbach, personal communication).

Step-by-Step Procedure

Preparation of Chicken Liver Nuclei

> Perfuse the livers with ice-cold 0.15 M NaCl solution (livers should be on ice).

> Put livers in an ice-cold beaker and add 4 volumes (volume : weight) of 2.2 M sucrose buffer.

> Mince the livers with scissors and homogenize in the cold with 4–5 strokes at 800 rpm in a glass-Teflon homogenizer. The homogenate should have a final concentration of 1.6–1.62 M sucrose.

> Overlay the homogenate on 4 ml of 1.8 M sucrose buffer in a SW 27 polyallomer centrifuge tube and centrifuge for 1–1.5 hours at 25 000 rpm in a SW 27 Beckman rotor at 4° C. Under these conditions the membranes remain at the top of the tube and the nuclei will sediment to the bottom of the tube.

> Remove the thick top layer of fat and membranes with a spatula, decant the supernatant, wipe the sides of the tube clean with a tissue and put the tubes on ice.

> For storage the purified nuclei are resuspended in a small volume of 50 mM HEPES, pH 7.5; 25 % glycerol; 2.6 % BSA; 0.1 mM EDTA; 0.15 mM spermine; 0.5 mM spermidine; 2 mM dithiothreitol. Adjust the volume of buffer to give about 10^9 nuclei per ml and store them in small aliquots in liquid nitrogen. For immediate DNA isolation, resuspend the nuclear sediment in a small volume of cold 0.15 M NaCl; 5 mM EDTA (final concentration of nuclei about 10^9 nuclei per ml).

Isolation of Nuclei from Cells in Tissue Culture

> Resuspend the cell pellet in 5–7 volumes (volume : weight) of 0.35 M sucrose buffer.

> Homogenize at 0° C with 20 strokes in a Dounce glass-glass homogenizer.

> Sediment the nuclei at 800–1000 × g for 10 minutes in a HB-4 Sorvall rotor.

> Decant supernatant fraction, resuspend the nuclei in the above buffer and centrifuge the crude nuclei preparation again.

> If the DNA is to be immediately extracted, resuspend carefully the nuclei in ice-cold 0.15 M NaCl; 5 mM EDTA.

DNA Extraction from the Nuclei

Note: *The DNA extraction procedure follows essentially the same protocol irrespective of whether nuclei from tissue cultures or tissues are used.*

> Add one volume of 20 mM Tris, pH 8; 20 mM NaCl, 20 mM EDTA; 1 % SDS containing 600 μg of proteinase K per ml to the nuclear suspension.

> Incubate at 37° C for 3 hours.

> Digest nuclear RNA by adding 50–100 μg of pancreatic ribonuclease A per ml; continue incubation at 37° C for 1–2 hours. At pH 8 in the presence of 0.5 % SDS pancreatic ribonuclease A is still active (Mendelsohn and Young, 1978) but after a short time (sufficient to destroy RNA) the enzyme will be destroyed by the proteinase K. At the end of incubation the very viscous mass of DNA is extracted at room temperature with an equal volume of phenol saturated with 1 M Tris, pH 8. The extraction can be done either by inversion in a closed Corex tube or in an Erlenmeyer flask in a rotating (giratory) shaker. The speed should be chosen so that the two

phases are mixed well. After 5 minutes of extraction, add one volume of chloroform, mix and separate phases in a Corex tube in a clinical centrifuge (a few minutes). Decant the viscous mass of DNA and re-extract 6–7 times as outlined above. Do not use any pipette during these extractions. A broad-tipped pipette may however be used in the last step of purification where the DNA preparation is extracted once with pure chloroform and the phases are separated by centrifugation at 7000 rpm in a Sorvall HB-4 rotor.

> After extraction the DNA can be either extensively dialyzed (1) or dialyzed and ethanol precipitated (2):

Test the quality of the isolated genomic DNA as follows:

(a) Electrophoresis of 1–2 μg of DNA on 0.8–1 % agarose gels. Staining with ethidium bromide will show whether the DNA fragments are large enough and without degradation. If the DNA is partially or completely degraded (i. e. on agarose gel presenting a smear) do not use for genomic sequencing.

(b) A digestion of a few μg of DNA with an appropriate restriction enzyme should indicate whether the DNA is sufficiently clean. If the DNA is not digested by a tested batch of restriction enzyme, then the enzyme could be inactive for one or more of the following reasons. Either the proteinase K was not completely removed and inactivated the restriction enzyme, or too many proteins were contaminating the DNA, or the dialysis was not complete and traces of SDS present in the original preparation inhibited the enzyme. At any rate, regardless of the cause of the inhibition of the restriction enzyme, it is best to repeat the above purification procedure once more.

(1) Dialyze in the cold (4°C) for 2–3 days with several changes of a large volume of 10 mM Tris, pH 8; 1 mM EDTA. The very viscous DNA will have about 1 mg DNA/ml and can be kept at 0°C in the presence of one drop of chloroform. Such preparations have been kept up to one year without any degradation. Alternatively it is possible to freeze the DNA preparation in small aliquots at $-80°C$. Frozen this way the DNA will remain intact for several years. However once a sample is thawed do not freeze it again. In order to speed up the dialysis it is possible to increase the porosity of the dialysis membrane by a treatment with $ZnCl_2$ (Craig, 1967): An aqueous solution of 64% zinc chloride is prepared. Fill up a membrane sack (previously treated as described above) with the solution of $ZnCl_2$ and dip it inside a cylinder containing the same solution. Let it stand for 1 to 2 hours at room temperature. Empty the dialyzing sack and cylinder and replace the solution inside and outside the bag by 0.01 M HCl. Let it stand for 1 to 2 hours at room temperature. Empty the dialysis sack and wash it thoroughly with water until the pH is neutral. Store the treated dialysis sack as described above.

(2) It is also possible to precipitate the DNA after 24 hours dialysis. Add sodium acetate to a final concentration 0.3 M, pH 5 and add 2.5 volumes of cold ethanol. After 2 hours at $-20°C$ or 30 minutes at $-80°C$, centrifuge the DNA at $16\,000 \times g$ for 10 minutes. Decant the ethanol and remove the remaining traces of solution with a stream of N_2 (or vacuo). Redissolve the DNA sediment with an appropriate volume of 10 mM Tris, pH 8. The volume should be chosen to give a concentration of 1 mg DNA/ml. Solubilization of the DNA will take about 24 hours at room temperature.

Notes:

ISOLATION OF GENOMIC DNA

2 Restriction Digest of Genomic DNA

The Choice of the Restriction Enzyme

One of the most critical steps of the genomic sequencing procedure is the choice of the appropriate restriction endonuclease. This enzyme is used to completely digest the genomic DNA to create a population of DNA fragments, a subset of which contains the target sequence. Good results will be obtained only if the DNA is completely and cleanly digested. It is therefore crucial that there is no nonspecific reaction by either the restriction enzyme (e. g. EcoRI giving EcoRI* digestion) or contaminating nucleases. The most suitable enzymes are those which yield DNA-restriction fragments of at least 900 base pairs. The longest used was over 2000 base pairs in length. Such fragments are still easily soluble in very small volumes of buffer, used for the loading of the sample onto the sequencing gel. In an optimal genomic sequencing experiment the sites of interest should be situated at least 50 base pairs and at the most 270 base pairs from one end of the target restriction fragment (Fig. 3).

The Restriction Digest

We found that the best results were obtained when diluted genomic DNA was digested overnight in small aliquots. The incubation buffers were always prepared according to the suppliers recommendations with the following modification: we did not use bovine serum albumin in order to avoid any possible DNAse contamination that may increase the number of nonspecific nicks within the DNA. The sterile-filtered and ten-fold concentrated incubation buffers were stored frozen in aliquots. The genomic DNA was digested overnight at the temperature recommended by the supplier and with a three-fold excess (units/μg DNA) of restriction enzymes (6 base-pair cutters) and a ten-fold excess (units/μg DNA) for 4 base-pair cutters. Following digestion, the genomic DNA should be treated at least once by phenol extraction.

After the digestion with restriction enzymes it is routine to perform at least one classical Southern-blot analysis (Southern, 1975) to check that 1) the DNA was digested to completion and 2) that there is only one DNA fragment, hybridizing with the radioactively labeled probe. If the DNA is only partially digested, a further digestion with the same enzyme is required, otherwise the backround within the sequencing lanes will be unacceptably high. In the case of excessive digestion (due to the presence of traces of contaminating nucleases in the enzyme preparation) with the restriction enzyme, the DNA should also not be used. In the case that more than one sharp band appears on the Southern-blot each restriction fragment should be isolated by means of preparative agarose gels. The precise area to be cut out from the gel is identified by Southern-blot analysis of a longitudinal strip of the gel. In special cases the same preparative procedure can be useful for an enrichment of the target fragment. The amount of DNA to be digested for genomic sequencing will depend on the size of the genome and the reiteration frequency of the gene to be studied. For example, a single copy gene from a genome of 2×10^6 kb (avian) will require for each lane on the sequencing gel 25 μg of genomic DNA to give a strong hybridization signal. Table 1 compiled from the literature gives an idea about the relative complexity of different genomes, hence the amount of DNA to be digested. However, one should bear in mind that irrespective of the size of the genome, always the same quantity of DNA has to be taken (number of cuts per 1000 nucleotides) for each chemical sequencing reaction.

▶

Fig. 3. Example of a highly resolved genomic sequence

A 1-m long sequencing gel was cut into three pieces, transferred to nylon membranes and hybridized with a 120-nucleotide-long single-stranded DNA probe. The lower part of the gel (right-hand panel) gave a resolution between nucleotide position (50) and (86) from the end of the restriction fragment, the middle part (middle panel) between nucleotide position (87) and (171) and the upper part (left-hand panel) between nucleotide position (172) from the end of the restriction fragment and the exclusion size of this 8 % gel which is approximately 300 nucleotides. The sequence shows a comparison of the genomic sequences (upper strand; nucleotide -460 to -680) of avian vitellogenin II gene from the liver (lanes 2 and 3) and erythrocyte DNA (lane 4) of egg-laying hens. The sites of methylation-demethylation of CpG at positions 525, 587, 612, 614 are marked by the arrowheads a–d, respectively; 1 represents the sequencing lanes of cloned vitellogenin gene (Saluz and Jost, 1986).

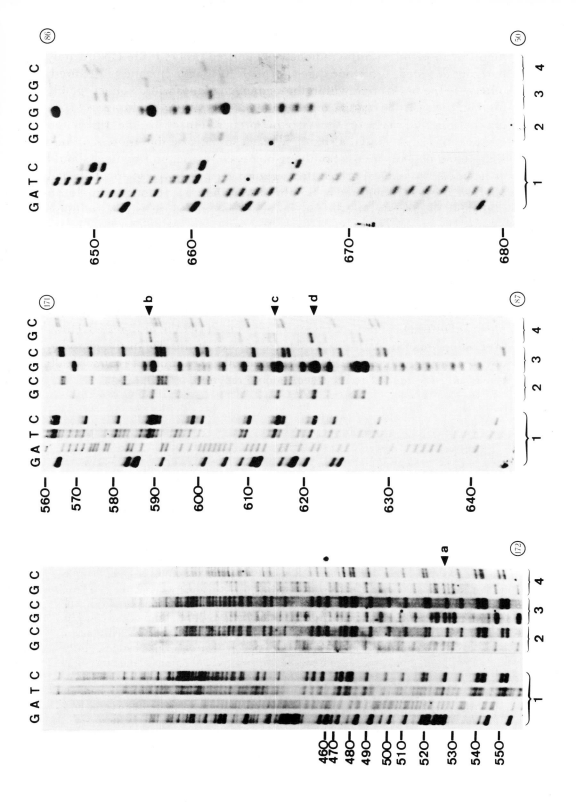

Table 1: DNA Content of Various Cells

Organism	Number of Base Pairs	Relative Complexity
* *E. coli*	4×10^6	1
* Yeast	1.35×10^7	3.3
* *Drosophila*	1.65×10^8	41
Chicken	2×10^9	500
Mammal	5×10^9	1 250
Maize	1.5×10^{10}	3 750
Lily	3×10^{11}	75 000

* Darnell et al., 1987

Materials and Buffers

> Spectrophotometer (260 nm)

> Vortexer

> Scissors or razor blade

> Genomic DNA

> Restriction enzyme of choice

> Water bath

> Incubation rack

> Eppendorf tubes (ev. silicon-treated)

> Clinical centrifuge

> Corex tubes

> Ultracentrifuge

> SW 27-rotor, Beckman or equivalent

> Ultracentrifuge tubes (polyallomer)

> Phenol (+ 0.1 % hydroxyquinoline; saturated with 1 M Tris, pH 8)

> Chloroform

> Sterile distilled water

> 10 × restriction buffer as recommended by the manufacturer (without bovine serum albumin)

> 3 M sodium acetate/0.005 M EDTA (pH 5)

> Ethanol

Step-by-Step Procedure

> Determine the OD_{260} of the purified genomic DNA: thaw a stock of genomic DNA and take an aliquot using a sterile capillary or a pipetman and tip with a broad end (cut the end with razor blade or scissors). Add the DNA to 1 ml of water. Mix well with a Vortexer and measure the OD_{260} with a spectrophotometer $1\,\mu g$ DNA $= 0.02\ OD_{260}$). Alternatively DNA concentration can be measured by the diphenylamine reaction or DABA reaction (see Appendix).

> The restriction digestion should be made in aliquots of $15\,\mu g$ each of DNA per $300\,\mu l$ of incubation mixture. Calculate the total number of aliquots needed for the digestion of genomic DNA. At least $50\,\mu g$ of digested genomic DNA is needed for one specific base reaction.

> To each silicone-treated Eppendorf tube (1.5 ml) add $30\,\mu l$ of a $10\times$ restriction buffer (use the same buffer recommended by the manufacturer but without bovine serum albumin; the sterile filtered $10\times$ buffer is stored at $-20°C$). Add the $15\,\mu g$ of genomic DNA, mix gently by tapping the tube. Add sterile water up to a final volume of $300\,\mu l$ per tube. Mix gently by tapping the tube and add 45 units of the chosen restriction enzyme. Should restriction enzymes with 4 bp recognition sequences be used, 150 units will be needed. Mix again as described above and give a short spin of a few seconds in a microfuge.

> Incubate overnight at the temperature recommended by the manufacturer.

> Pool the aliquots of digested DNA (usually 10–15 aliquots) in a 15-ml silicone-treated, sterile Corex centrifuge tube. Add 0.5 volume (v/v) of distilled phenol (+ 0.1 % hydroxyquinoline) saturated with 1 M Tris, pH 8. Cap the Corex tube and mix several times by inversion. Add the same volume of chloroform and mix again as described above. Centrifuge in a clinical centrifuge (3000 × g for 5 minutes).

> Transfer the aqueous phase with a sterile pipette (without touching the interphase) into a SW-40 or SW-27 polyallomer Beckman centrifuge tube.

> Add 1/10 volume of 3 M sodium acetate, 0.005 M EDTA (pH 5) and add 2.5 volumes of ethanol. Mix well by inversion and leave it overnight at −20° C.

> Centrifuge 25 000–30 000 rpm, 4° C, 1 hour), decant the supernatant and dry the DNA sediment under vacuum. Dissolve the DNA in 200 μl of sterile distilled water.

> Determine the OD_{260} from an aliquot and store the DNA at −70° C if it is not to be used immediately for the Maxam & Gilbert sequencing reactions.

Notes:

Notes:

3 Chemical Sequencing Reactions on Restricted DNA

The chemically induced cleavage at the bases G, A+G, T+C and C described below is based on a slight modification of the original Maxam & Gilbert reactions (1980). For the T-reaction we used the method of Rubin & Schmid (1980). The three principal steps in all chemical sequencing reactions are essentially the same: modification of the base, removal of this base from its sugar and a piperidine induced cleavage at this position. The first step is base specific, random and limited. The beta-elimination step is quantitative.

The thymine-specific reaction (Fig. 4) is necessary only when studying a strand-specific methylation pattern. 5-methyl cytosine can be distinguished from cytosine by its lack of reaction with hydrazine resulting in the disappearance of a band in the C-specific sequencing lane. To ascertain that the absence of a band in the sequence represents 5-methyl cytosine and not thymine (arising by a deamination of 5-mC), thymine-specific sequencing reactions have to be performed. Several different thymine-specific reactions are available (Friedmann and Brown, 1978; Rubin and Schmid, 1980; Saito et al., 1984). The most convenient procedure, based on potassium permanganate oxidation of thymidine, was described by Rubin and Schmid (1980).

One of the major differences between the classical Maxam & Gilbert sequencing procedures and genomic sequencing is the amount of DNA used in each chemical reaction. In genomic sequencing reactions a much higher concentration of unlabeled DNA is required to effectively reduce the number of chemical modifications (therefore also beta-eliminations $\triangleq$ number of 'cuts') per unit length of DNA. Reducing the number of cuts by decreasing the temperature of the reaction or by diluting the reagent is also possible but more problematic. The number of cuts per target DNA molecule can greatly affect the resolution of the genomic sequence. If at least one of the two cuts is located within the probe region, for example, two or three hybridizable fragments are created. Such fragments result in additional unspecific bands within the sequencing lanes and makes the sequence more difficult if not impossible to interpret (see 'Trouble-Shooting Guide': Fig. 21/2).

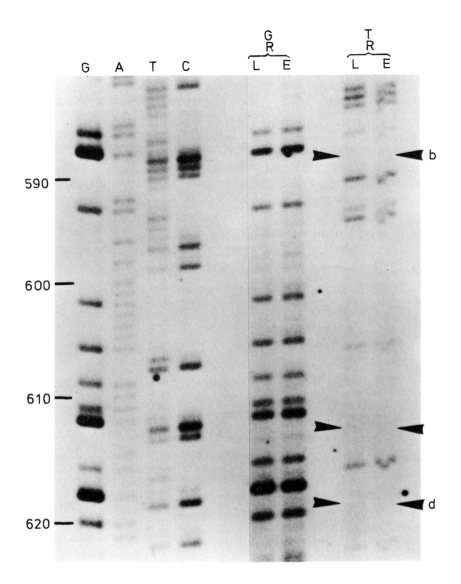

Fig. 4. T-specific reaction

Comparison of G- and T-specific reactions of potential methylation sites (b, c, d) in the upstream area (upper strand) of the avian vitellogenin gene II (L = liver DNA, E = erythrocyte DNA. R stands for rooster). In this case no deamination of 5-mCytosine could be detected. At the left 4 control reactions (G, A, T, C) performed as described in the text. (Saluz et al., 1986)

SEQUENCING REACTIONS ON RESTRICTED DNA

Finally it is of a paramount importance that an excellent vacuum be used to eliminate piperidine and that the number of lyophilizations described in the 'Step-by-Step Procedure' is followed. Any trace of piperidine typically results in the smearing of bands during electrophoresis on sequencing gels.

Control Reactions with Cloned Plasmid DNA

The control reactions with cloned DNA are important for two reasons: 1) they allow a precise orientation and interpretation of the genomic sequencing lanes and 2) they are used as an internal standard for the hybridization step (see chapter III.7).

The Chemical Reagents

Most of the chemicals used for the sequencing reactions are labile and toxic:

> Hydrazine is very unstable. Its oxidation results in diimine. The use of partially oxidized hydrazine results in side reactions, especially with thymidine. Hydrazine should thus be kept only under nitrogen, in a dark bottle and in the cold (cold room) or frozen in aliquots ($-20°$ C). The working solutions of hydrazine should be replaced daily.

> Dimethylsulfate (DMS) hydrolyzes to sulfuric acid and methanol if it picks up moisture from the air (dimethylsulfate is anhydrous). Therefore the bottles containings DMS should always be well closed and the DMS should be stored under nitrogen.

> Potassium permanganate ($KMnO_4$): The working solution should always be freshly prepared, as it oxidizes very rapidly when exposed to air. It is not advisable to keep frozen stock solutions.

> Allyl alcohol should be kept in a cold and dark place. After a few weeks the bottle has to be replaced.

> Piperidine undergoes oxidation when brought into in contact with air. Therefore it should be stored under nitrogen.

> All the buffers used for the sequencing reactions are frozen in aliquots at $-20°$ C. They are opened only once and discarded after use.

Note: *The bottles containing hydrazine, piperidine, dimethylsulfate should be stored, opened and handled only in a well-ventilated fumehood. Residues of hydrazine are detoxified in 3 M ferric chloride and dimethylsulfate in 5 M sodium hydroxide solution. Only chemically trained and experienced people should consider distilling the chemicals described above.*

Materials and Buffers

> Speed vac (Savant)

> Eppendorf centrifuge

> Sorvall or equivalent centrifuge

> SS-34 Sorvall rotor or any equivalent rotor with adapters for Eppendorf tubes

> Dimethylsulfate (DMS) of the highest purity (Aldrich)

> Dimethylsulfate (DMS) buffer: 50 mM sodium cacodylate, pH 8; 1 mM EDTA.

> Dimethylsulfate stop buffer (DMS stop): 1.5 M sodium acetate (NaOAc) pH 7; 1.0 M mercaptoethanol

> Formic acid (p. a.; Fluka)

> Hydrazine (HZ) of the highest purity (Aldrich)

> Hydrazine stop buffer (HZ stop): 0.3 M sodium acetate pH 7.5; 0.1 mM EDTA

> 0.3 M sodium acetate (NaOAc), 0.5 mM EDTA, pH 5

> 1.27×10^{-4} M potassium permanganate (KmNO$_4$); freshly prepared

> Allyl alcohol (purum; Fluka)

> Ethanol

> Dry ice

> Piperidine of the highest purity (Fisher Scientific Company)

> Sample dye: 94 % formamide, 10 mM Na$_2$EDTA (pH 7.2); 0.05 % xylenecyanol, 0.05 % bromophenol blue (BPB)

Step-by-Step Procedure

Chemical Reactions on Control DNA

The control reactions were also carried out as described below except that genomic DNA was replaced by cloned DNA, containing the target sequence (in pBR 322; pBR 322 + insert app. 9 kb). The plasmid DNA was first digested with the same restriction enzyme used for the genomic DNA. For the G or C reactions $4\,\mu g$ and for the (G+A) or (T+C) reactions $8\,\mu g$ of the digested cloned DNA were mixed with bacterial DNA to give a total of $50\,\mu g$. After the final Maxam & Gilbert reactions the control DNA was dissolved in $300\,\mu l$ of water and frozen at $-70°\,C$ in $5-10\,\mu l$ aliquots. As controls for each genomic sequencing gel, $1-2\,\mu l$ of that reaction mixture was used per control lane.

G-reaction:

> Dry $50-75\,\mu g$ of digested genomic DNA in the speed vac.

> Dissolve the sediment of DNA in $6\,\mu l$ of water and $200\,\mu l$ of DMS buffer.

> Chill sample on ice and add $1\,\mu l$ of DMS.

> Mix by tapping the tube and centrifuge a few seconds in a microfuge (in the cold room).

> Incubate sample in a water bath for 10 minutes at $20°\,C$.

> At the end of incubation add $50\,\mu l$ of DMS stop buffer, mix and add $750\,\mu l$ of precooled $(-20°\,C)$ ethanol.

> (✱ ✱ ✱)
  Mix well by inversion and chill for 15 minutes in a
  mixture of dry ice and ethanol ($-70°$ C).

> Centrifuge tubes for 20 minutes in a SS-34 Sorvall
  rotor at 15 000 rpm at $0°$ C.

> Pour out the supernatant very carefully and cen-
  trifuge again for a few minutes in the microfuge.

> Remove the residual ethanol with a drawn out
  glass capillary.

> Resuspend the pellet in $250\,\mu$l of sodium acetate/
  EDTA, pH 5 (0.3 M/0.5 mM).

> Add $750\,\mu$l of precooled ethanol, mix thoroughly
  by inversion, chill at $-70°$ C and centrifuge as
  described above.

> To wash the pellet, add carefully 1 ml of 70%
  ethanol/water, centrifuge for 5 minutes and carefully pour out the supernatant and centrifuge
  again for 1 minute.

> Remove the residual ethanol with a drawn out
  glass capillary and dry the pellet in the speed vac.

> The DNA is ready for piperidine treatment (see at
  the end of the chapter).

(G+A)-reaction:

> Dry 50–75 μg of digested genomic DNA in the speed vac and dissolve pellet in 11 μl of water by tapping the tube.

> Chill on ice and add 25 μl of formic acid (concentrated).

> Mix by tapping the tube and centrifuge for a few seconds in a microfuge.

> Incubate in a water bath at 20°C for 4.5 minutes.

> Add consecutively 200 μl of hydrazine stop buffer and 750 μl of cold ethanol.

> Continue as described for the G-reaction (✱✱✱).

(T+C)-reaction:

> Dry 50–75 μg of digested genomic DNA in the speed vac and dissolve DNA pellet in 21 μl of water.

> Chill on ice and add 30 μl of hydrazine.

> Mix by tapping the tube and centrifuge for a few seconds in a microfuge.

> Incubate sample in a water bath at 20°C for 10 minutes.

> Add consecutively 200 μl of HZ stop buffer and 750 μl of cold ethanol.

> Continue as described for G-reaction (✱✱✱).

C-reaction:

> Dry 50–75 μg of digested genomic DNA in the speed vac.

> Dissolve pellet in 5 μl of water and add 15 μl of 5 M sodium chloride (NaCl).

> Mix by tapping the tube and chill on ice.

> Add 30 μl of hydrazine.

> Mix by tapping the tube and centrifuge a few seconds in a microfuge.

> Incubate sample at 20° C for 10 minutes.

> Add consecutively 200 μl of hydrazine stop buffer and 750 μl of cold ethanol.

> Continue as described for G-reaction (✽✽✽).

T-reaction:

> Dry 50–75 μg of digested genomic DNA in the speed vac.

> Dissolve pellet in 5 μl of water.

> Denature DNA at 90° C for 2 minutes; then quick chill in ice/water.

> Add 20 μl of 1.27×10^{-4} M potassium permangnate (KMnO$_4$).

> Mix by tapping the tube and centrifuge a few seconds in a microfuge.

> Incubate in a water bath of 20° C for 10 minutes (until the mixture is pink).

> Stop the reaction with 10 μl of allyl alcohol, freeze quickly at −70° C and lyophilize in a speed vac.

> The DNA is ready for beta eliminations (see below).

Beta-eliminations: *This step is the same for all the different reactions, and its purpose is to introduce a strand break at the modified bases.*

> Dissolve pellets in $100\,\mu$l of 1 M piperidine ($100\,\mu$l piperidine and $900\,\mu$l of water; freshly prepared).

> Incubate in a water bath at $90-95°$C for 30 minutes.

> Freeze sample at $-70°$C and lyophilize under a high vacuum (0.01–0.001 TORR).

> Dissolve pellet in $100\,\mu$l of water, freeze and lyophilize.

> Repeat last step at least twice and dissolve the pellets in $20\,\mu$l of water, divide into two aliquots of $10\,\mu$l each. Samples can now be stored at $-80°$C until required.

> Lyophilize one aliquot and dissolve it into $2\,\mu$l of water, add $5\,\mu$l of sample dye (see 'Materials and Buffers'; if possible dissolve the pellet in $1\,\mu$l of water and $3\,\mu$l of sample dye), heat for 1–2 minutes at $95°$C, chill in ice/water and load the samples onto the gel.

> **Control Samples:** *Take 1–2$\,\mu$l of specific reaction product per control sequencing lane and add 25$\,\mu$g E. coli DNA (in water). Dry the sample in the speed vac and dissolve pellet in water and sample dye as described above for genomic DNA.*

Notes:

4 Separation of Reaction Products on a Sequencing Gel

Gel Electrophoresis

A mixture of millions of different fragments of genomic DNA produced by the restriction-endonuclease digest and the chemical sequencing reactions must be separated by size on polyacrylamide gels. Therefore a very high standard is required when preparing and running the sequencing gel. Since nonspecific nicks within the target sequence increase the background and reduce the resolution of the sequence, it is advisable to use 1-m-long gels or more. The separations on such gels will lead to a greater distance between the bands so that the background, due to the nonspecific degradation products or hybridization-mismatches, will be diluted. For the gel system described in this book an electrophoresis of about 11 hours is optimal for a maximal resolution of the DNA sequence. For an electrophoresis of 11 hours or more a careful choice of the ionic strength of the buffer is required. If its buffering capacity is too low, undesirable changes in the pH (electrolysis of the buffer

anode: $H_2O \rightarrow 2e^- + 2H^+ + \frac{1}{2}O_2 \uparrow$; pH $\downarrow$;
cathode: $2e^- + 2H_2O \rightarrow 2OH^- + H_2 \uparrow$; pH $\uparrow$)

will result and decrease the resolution of the DNA sequence. On the other hand increasing the ionic strength decreases the velocity of the migrating DNA fragments within the electrical field. If you try to compensate by using a higher current, overheating will affect the separation by causing a trailing of the sequence ladder. These problems can be overcome by using either large volumes of buffer or changing the buffer at regular intervals. Best results were obtained with Tris-borate-EDTA (TBE). The best resolution of the genomic sequence (up to 300 bases) was obtained on an 8 % polyacrylamide gel with a ratio of 29 : 1 of acrylamide : bisacrylamide. The concentration of the chemical catalysts was chosen such as to give a short polymerization time preventing the formation of undesirable polyacrylamide gradients within the gel. During the casting of the gel two glass plates plus spacers are placed in a slanted position (see Fig. 5). The same position is maintained during the polymerization. To ensure complete polymerization the gel should be left for at least 12 hours before use. Prerunning the gel overnight at low current (11 mA) removes all charged impurities that may interfere with the perfect separation of the genomic DNA fragments. To avoid overheating of the upper end of the gel, the glass plates should not be in direct contact with the upper buffer chamber. Therefore we use a bridge, made of Whatman 3 paper (see Fig. 6), between the buffer tank and the gel. It is also important that the buffer level in the upper chamber be about 0.5 cm above the top of the glass plates, because if the bridge were to dry out it could easily ignite. In recent experiments, as an additional safety precaution, we incorporated an overheating cut-out device as part of the gel assembly. This consists of an electronic sensor that automatically switches off the power supply when the temperature of the outer surface of the glass plate exceeds 65° C.

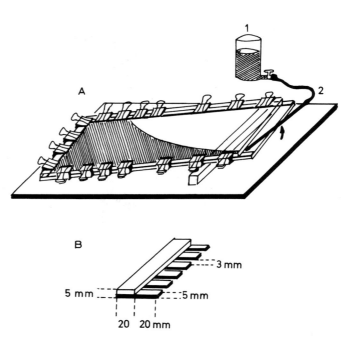

Fig. 5. Casting of the gel

During pouring of acrylamide the two assembled glass plates are placed in a slanted position (*A*) which is maintained during the polymerization. A vessel (1) connected to a tube and a 1-ml pipette (2) facilitates the pouring of the acrylamide solution. The teeth of the comb (*B*) should be 20 to 25 mm long and have a tight fit. The distance between two teeth is about 3 mm.

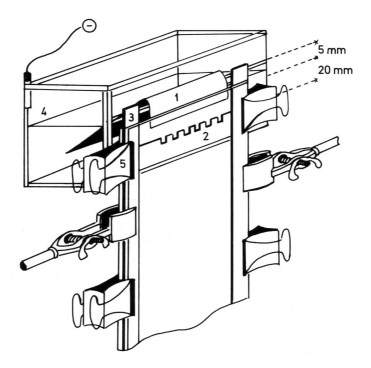

Fig. 6. Setting up the gel for electrophoresis

To avoid an overheating of the upper part of the gel, the glass plates should not be in direct contact with the upper buffer chamber. Therefore a space of about 2 cm is maintained between the glass plates and the upper buffer tank. The gel is loaded with the samples and Whatman 3 paper bridge (1) is placed between the buffer tank and the gel. The buffer level (4) is about 0.5 cm above the top of the glass plates. The wells (2) have to be cleaned before loading the gel. To avoid leakage of the buffer, the spacers (3) should be fixed with several metal clamps (5).

Materials and Buffers

> Power supply (at least 2500–3000 V and 100–200 mA)

> Glass plates: 1000 mm × 330 mm × 5 mm

> Spacers (PVC): 2 × (1 mm × 20 mm × 1000 mm);
  1 × (1 mm × 20 mm × 40 mm).

> Comb (PVC): 15 teeth (1 mm × 5 mm × 25 mm;
  space between the teeth: 3 mm)

> Vessel for pouring the gels according to Fig. 5 (volume: 400 ml)

> Filter paper (Schleicher & Schuell, No. Ls 14$^{1/2}$)

> Buechner flask

> Capillaries (100 μl)

> Saran wrap

> Middle viscosity silicone grease

> 1 % dimethyldichlorosilane in carbon tetrachloride

> Acrylamide (recrystallized 2 ×)

> Bisacrylamide (recrystallized 2 ×)

> Ammonium persulfate (analytical grade)

> Urea (analytical grade)

> N,N,N′,N′-Tetramethylenediamine (TEMED; analytical grade)

> Tris-borate-EDTA buffer (TBE) – 10 × TBE: 0.89 M Tris base,
  0.89 M boric acid, 0.02 M EDTA, pH 8.3

Step-by-Step Procedure

> Clean the glass plates with water and ethanol (when needed acetone to remove the residual dimethyldichlorosilane).

> Treat both plates with 1% dimethyldichlorosilane in carbon tetrachloride in the fumehood.

> Clean the plates again with ethanol and water.

> Fix the spacers to the bottom plate by means of a small trace of middle viscosity silicone grease. One spacer is placed at the end of the plate (bottom of the gel) and two other spacers are placed along the two sides of the plate.

> Complete the 'sandwich' with the upper plate using silicone grease as above.

> Fix the two plates on the spacer with clamps.

> **Attention:** *The clamps should only be in contact with the glass plates on the spacer. To avoid any leakage of the polyacrylamide, the clamps should be placed on the bottom part and lower part of the gel close to each other (see Fig. 5).*

> Pour the polyacrylamide solution as indicated in Fig. 5. The plates placed in a slightly slanted position should be filled (see arrow) along one edge only.

8% Polyacrylamide gel:

> For a gel of the dimensions given above you should prepare 400 ml of polyacrylamide solution.

> 30.92 g acrylamide (recrystallized twice)

> 1.08 g bisacrylamide (recrystallized twice)

> 168 g urea (analytical grade)

> 40 ml of 10 x TBE buffer

> Distilled water to 400 ml

> Dissolve the ingredients into solution by mixing with a magnetic stirring bar.

> Add 5.2 ml of 10 % freshly prepared ammonium persulfate and mix well.

> Filter the solution through Schleicher & Schuell, No. Ls $14^{1/2}$ filter paper.

> Degas the solution in a Buechner flask under vacuum for at least 5 minutes (shake bottle gently from time to time).

> Add 80 μl of TEMED.

> Mix gently.

> Pour the gel as indicated in Fig. 5 avoiding air bubbles. Insert the comb on top of the gel taking care not to trap air bubbles. Insert the comb 2–2.5 cm deep in order to have enough space for the insertion of the paper bridge. Under the conditions described above the polymerization takes about 10 minutes at room temperature.

> After polymerization cover the top of the gel with several layers of wet paper towels and with Saran wrap. To ensure complete polymerization store the gel overnight at room temperature.

Preelectrophoresis and electrophoresis:

> To position the paper bridge (shown in Fig. 6) correctly, it is necessary to cut away the top of the gel to a depth of approximatively 1 cm. This is best done in small segments, using a bent 21 gauge (0.8 mm × 40 mm) syringe needle.

> Set up the gel as indicated in Fig. 6.

Separation of Reaction Products

> Use a Whatman 3 paper bridge between the upper buffer tank and the gel. The buffer is 1× TBE.

> Use a needle and a syringe filled with buffer and remove the air bubbles on the bottom of the gel. The preelectrophoresis is performed overnight with a constant current of 11 mA (approximately 300 V).

> Change the buffer in the upper and lower chamber.

> Heat up the gel for about 1 hour at constant current (50 mA). The surface of the gel will reach approximately 45° C.

> Switch off, disconnect the bridge (put it into the upper buffer chamber to ensure it does not dry out) and clean the slots with a stream of 1 × TBE, using a syringe.

> Load the samples very carefully into the wells of the gel (with long drawn out capillary tubes; the capillary part is approximately 3 cm long;

  Note: *Do not load all of the sample; a very small amount should be left in the capillary tip to avoid introducing air bubbles into the wells).*

> Reconnect the gel to the tank carefully with the paper bridge avoiding any turbulence.

> Switch the power on and increase the current to 60 mA.

> If the volume of the chamber is 750 ml, it is necessary to change the buffer every 6 hours. For an overnight electrophoresis of 11 hours use chambers with sufficient capacity, i. e. 2000 ml.

> Under the conditions we described here electrophoresis is continued until the xylenecyanol marker migrated 68−76 cm from the top of the gel, i. e. 11 hours at 58−62 mA (constant current).

Notes:

5 Electrotransfer to Nylon Membranes

One of the most important and technically difficult aspects of genomic sequencing is the electrophoretic transfer of the genomic DNA from the sequencing gel to the nylon membrane. Many different parameters have to be taken into consideration: electrical resistance, strength of the electric field, diffusion, Joule heating, stability of the buffer, etc. Most of the transfer systems described in the literature so far (Towbin et al., 1979; Gibson, 1981; McCellan and Ramshaw, 1981; Shuttleworth, 1984) do not fulfill the optimal conditions for genomic sequencing. A high resolution electrotransfer of the sequencing ladder requires an absolutely tight contact between the gel and the nylon membrane. Trapped gas bubbles (arising from electrolysis of the buffer) in the transfer system disturb the electric field and result in the distortion or loss of DNA bands in the blot. The electrical field has thus to be low enough to avoid buffer electrolysis, while at the same time being sufficiently high to ensure a quantitative diffusion-free transfer of the DNA. The ionic strength of the buffer has to be chosen in such a way that a constant pH

1 Weight
2 Electrode plate
   (anode)
3 Whatman-17 (2 cm)
4 Filter membrane
5 Gel
6 Buffer level
7 Whatman-17 (1 cm)
8 Electrode plate
   (cathode)
9 Teflon cube
10 Plastic box

Fig. 7. The electrophoretic transfer system as modified from Vaessen et al., 1981

In a plastic box place the lower sieve plate electrode (cathode) on four Teflon cubes. Then add a first layer of Whatman 17 paper, the sequencing gel, the immobilizing membrane matrix, a second layer of Whatman 17 paper and the upper sieve plate electrode (anode).

A weight placed on the top of the anode ensures a tight contact between the gel and the membrane (Saluz and Jost, 1986).

during the transfer time is guaranteed without perturbing the velocity of the DNA molecules to be transferred, and the volume of the buffer should be large enough so that its temperature is constant over the transfer time.

The best results were obtained with the system shown in Fig. 7. This modification of Vaessen's transfer system (Vaessen et al., 1981) is easily constructed, cheap and very efficient. It consists of a lower sieve plate electrode (cathode), a layer of Whatman 17 paper, the sequencing gel, a nylon membrane, a second layer of Whatman 17 paper and an upper sieve plate electrode (anode). The 'sandwich' is supported on four Teflon cubes in a plastic dish so as to permit gas bubbles to escape. A weight on top of the anode guarantees a tight contact between gel and membrane. The distance between the two electrodes is only about 3 cm so that with a high current of 1.2–2 A only about 32–35 volts are required. To avoid air bubbles between the different layers of the transfer system, the Whatman papers and the nylon matrix have to be soaked in the transfer buffer for at least 30 minutes and preferably longer. Air bubbles should be squeezed out of the paper while submerged in a large volume of buffer (1 × TBE).

Because of the high current used, safety precautions should be taken.

Materials and Buffers

> Power pack for high current (at least 2–3 A) and low voltage (appr. 60–100 V)

> Whatman 17 paper

> 2 stainless steel plate sieve electrodes (1.5 mm × 150 mm × 400 mm) (see Fig. 22)

> 4 Teflon cubes (0.5 cm)

> Weight of 2.1 kg (for example bottles with water)

> Plastic tank for transfer (415 mm × 210 mm × 100 mm)

> Plastic boxes for preincubation of membranes (415 mm × 160 mm × 50 mm)

> Gene Screen membranes

> 10 × Tris-borate-EDTA buffer (TBE): 0,89 M Tris base, 0.89 M boric acid, 0.02 M EDTA, pH 8.3

> Saran wrap

> Waterproof color marker

> Scalpel blade

Step-by-Step Procedure

Preparation of the Sequencing Gel for the Electrotransfer:

> At the end of the electrophoresis remove the gel from the electrophoresis apparatus and take the upper glass plate away by carefully opening the 'sandwich' at one end with a flat piece of metal. Cover the gel with Saran wrap and avoid trapping any air bubbles by pressing the Saran wrap with soft paper.

> Mark the pieces to be transferred (in our case an area of 15 × 30 cm) directly on the Saran wrap by means of a color marker. Cut out the pieces of the gel one after the other using a scalpel blade. If the gel remaining on the glass plate has also to be transferred take care that it does not dry out; cover the cut side with a small strip of Saran wrap.

Preparation of the Electrophoretic Transfer System:

> Cut a sheet of Gene Screen to the exact size of the gel to be transferred (wear gloves to handle the nylon membrane); cut one corner to aid subsequent orientation of the Gene Screen membrane.

> Float the membrane on 1 × TBE at room temperature for at least 10 minutes. Gently rock the plastic box so that the membrane sinks and is covered with buffer. After another 30 minutes the membrane is ready for the transfer.

> The assembly of the transfer system is illustrated in Fig. 8:

(1) Place the cathode electrode on four Teflon cubes positioned in the corner of a plastic dish (415 mm × 210 mm × 100 mm).

(2) Place 8 layers, one after the other, of wet Whatman 17 papers (30 × 16 cm) presoaked with 1 × TBE (Whatman papers were incubated in 1 × TBE and pressed from time to time to avoid trapped air bubbles) onto the cathode. Pour 800 ml (optimal buffer volume for plastic tank described above) of 1 × TBE onto Whatman papers. The buffer level in the plastic tank should not exceed the 3rd or 4th upper sheet of Whatman 17 paper. Rock plastic tank so that air bubbles can escape. Wait until there is no buffer remaining on top of the Whatman paper.

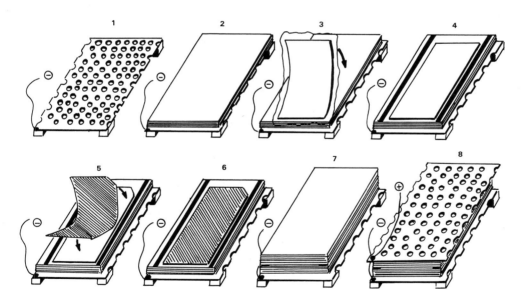

Fig. 8. Setting up the transfer system

The transfer system is built up as described in the 'Step-by-Step Procedure'.

(3/4) Place the gel, covered with the Saran wrap, on top of the filter papers (avoid air bubbles); to help visualize the bubbles use a spotlight placed at a low angle to the gel. Press the gel gently onto the Whatman layer to get rid of air bubbles by using a piece of soft paper. Peel off the Saran wrap carefully. Place along both sides of the gel spacers of the same thickness as for gel electrophoresis so that the upper and lower layer of Whatman papers are not in direct contact.

(5/6) Place the wet Gene Screen membrane onto the gel avoiding air bubbles (once the membrane has touched the gel it may no longer be removed).

(7) Complete the transfer system by carefully placing 16 sheets, one after the other, of Whatman 17 paper (also soaked with $1 \times$ TBE as described above) onto the membrane (papers should be just wet but not dripping with buffer).

Note: *As for the classical Southern-blots the upper layers of filter paper on top of the nylon membrane should never come into contact with the lower layer of filter paper.*

(8) The anode electrode is placed on top of the second Whatman layer; place a weight of 2 kg on top of it (bottle(s) with water will do the trick). Transfer DNA for 30–35 minutes at 1.2–1.8 A (about 32 V) at room temperature. In the meantime cover a glass plate tightly with Saran wrap.

> At the end of the transfer switch off the power supply and take the anode and the upper Whatman 17 layer away carefully (Whatman 17 papers are reused with the exception of the sheets just below the anode and the gel, respectively).

> Irradiate the wet membrane under UV light as described in the following chapter.

> If a second transfer follows, change the buffer and soak the Whatman 17 papers once more in 1 × TBE and repeat the above steps.

Notes:

6 Immobilization of DNA on Nylon Membranes

The Immobilizing Matrix

Two common types of membrane suitable for nucleic-acid hybridization are currently on the market: nylon and nitrocellulose filters. The nylon membranes, because of their greater physical strength are generally preferred to the nitrocellulose membranes. Genomic sequencing requires nylon membranes of highly homogeneous composition. However for technical reasons it seems that the manufacturers of nylon membranes are not yet able to produce membranes of the quality required for genomic sequencing. Unfortunately the quality of nylon membranes changes from batch to batch. Variation in the quality of membranes does not seem to be important in Southern or Northern-blotting. However where the limits of detection go as far as a few femtograms of nucleic acids, as is the case for genomic sequencing, any slight change in the local binding capacity of the filter will result in irregularities in the strength of the hybridization signals. It is then necessary to test each new batch of membrane. The necessary procedure is described at the end of this chapter (see 'Step-by-Step Procedure').

Binding of DNA to Filter Membranes

Three different types of membranes were tested for their ability to bind DNA stably: Gene Screen^{TM}, Zeta-Probe^{TM} and Millipore. Under appropriate conditions (variety of buffers, heat treatment, UV treatment or a combination of both) it was possible to bind nearly 100 % of the input DNA onto the different membranes. However, as we shall see, such a strong binding of DNA to the membranes is undesirable for molecular hybridization. The DNA binding capacity of a filter was determined as follows: total genomic DNA was labeled radioactively by nick-translation and then was submitted to one of the sequencing reactions of Maxam and Gilbert. The product of reaction was spotted on $0.2 \, cm^2$ pieces of filter. For the fixation of DNA to the filter UV irradiation was carried out as described in Fig. 9. The UV-flux (254 nm) at a distance of 22 cm was $0.39 \, mWatt/cm^2$. Filters that had been pretreated for 30 min in appropriate buffers were still humid during the UV-irradiation. Following UV-irradiation and/or heat treatment the radioactivity of the filterbound DNA was measured (value of input DNA on filter). The filter pieces were washed immediately using stringent washing conditions and the remaining radioactivity determined (stably bound DNA on filter). An example of such a titration is shown in Fig. 9. After 5 min UV irradiation, 50 % of the input DNA was covalently bound to the Gene Screen membrane. Other batches of membranes needed up to 20 min of irradiation for the same degree of DNA binding under the same conditions. In the next chapter we shall see that the percent of DNA binding to the filters plays a crucial role for the efficiency of subsequent hybridization. For this reason it is of paramount importance that each new batch of filter membrane be tested for its capacity to bind DNA.

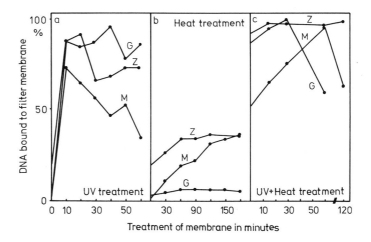

Fig. 9. Effect of UV, heat treatment or both on the stable binding of labeled denatured DNA to different membranes

For the tests the following types of membranes were used: Gene Screen (G), Zeta-Probe (Z) and Millipore (M). Before spotting the heat-denatured DNA, filters were pretreated for 30 min in $1 \times$ TBE for G and Z and in $20 \times$ SSC for M. UV irradiation (panel a) was done at a UV flux of $0.39 \, mW/cm^2$ at 254 nm. Heat treatment (panels b and c) was carried out in a vacuum oven at $86°$ C. For panel c the pretreatment with UV was 45 min for G and M, and 20 min for Z (Saluz and Jost, 1986).

Materials and Buffers

Testing the efficiency of DNA binding to nylon membranes:

> UV-chamber containing 2 UV tubes from Philips (TUV 15 W, G 15 T 8) (see Fig. 23 & 24)

> Vacuum oven

> Gene Screen™ membranes

> Plastic box (dimensions: $330 \times 220 \times 50$ mm)

> Glass plate

> Color marker

> Saran wrap

> Timer

> Water bath for heat denaturation

> Scintillation vials

> $1 \times$ Tris-borate-EDTA buffer ($1 \times$ TBE) – $10 \times$ TBE: 0.89 M Tris base, 0.89 M boric acid, 0.02 M EDTA, pH 8.3

> $0.1 \times$ SSC (prepared from stock $20 \times$ SSC: 3 M sodium chloride, 0.3 M sodium citrate, pH 7.4)

> Sodium dodecyl sulphate (powder, cryst. research grade)

Covalent binding of the genomic DNA fragments to the nylon membrane:

> UV-chamber, containing 2 UV tubes from Philips (TUV 15 W, G 15 T 8 or equivalent) (see Fig. 23 & 24)

> Vacuum oven

> Apparatus for heat and vacuum sealing

> Plastic bags for sealing

> Whatman 17 paper

> Nylon membrane

Step-by-Step Procedure

Testing the covalent binding of DNA to new batches of Gene Screen membranes:

The aim of this test is to measure the DNA binding capacity of new batches of filter upon UV irradiation. This procedure should be repeated with all new rolls of filter as well as after changing UV lamps.

> Prepare some total nick-translated genomic DNA (appr. 6×10^8 cpm/μg) and submit it to the C-reaction as described by Maxam & Gilbert (Maxam & Gilbert, 1980; Alternatively, oligonucleotides (app. 50-MER) labeled by the kinase reaction). After the final sequencing step dry this DNA in a speed vac and dissolve the pellet in water to a concentration of about 5×10^6 cpm/5 μl water. Keep the DNA on ice or in a freezer at $-20°$ C.

> Float a small strip (5 cm $\times$ 30 cm) of membrane on the surface of 1 $\times$ TBE buffer in a plastic box at room temperature for at least 10 minutes. Gently rock the plastic box until the filter sinks and leave it for 30 minutes at room temperature. In the meantime cover a glass plate tightly with Saran wrap (smooth surface).

> Place the wet membrane on a piece of parafilm, allow to dry partially and cut the filter into pieces of 1 cm^2.

> Pipette 5 μl of heat-denatured labeled DNA onto the center of each filter (denature DNA at 95° C for 1 min and then chill in ice/water).

> Place the moist filters with the DNA on the Saran wrap/glass plate.

IMMOBILIZATION OF DNA ON NYLON MEMBRANES

> Expose the filters to UV for 1, 3, 5, 8, 10, 13, 16, 20 and 25 minutes (in duplicates) and mark the approximate position of the glass plate under the UV-light using a color marker to ensure reproducibility in subsequent irradiation.

> Put each piece of irradiated membrane into separate 25 ml plastic scintillation vials and measure the radioactivity (cpm according to Cerenkov; Jelley J. V., 1958).

> Wash the filters directly in the scintillation vials by adding 10 ml of preincubated (65°C) stringent washing buffer (0.1 × SSC/0.5% SDS) and incubating the vials at 65°C in a shaking (80 rpm) water bath. After 30 minutes exchange the buffer and incubate for another 60−90 minutes.

> Remove buffer

> Count the radioactivity (Cerenkov counts) again and calculate the percentage of the binding:

$$\% \text{ Binding} = \frac{100 \times \text{cpm of bound DNA}}{\text{cpm of input DNA}}$$

> The maximum hybridization signal is obtained when roughly 30% to 50% of the DNA is stably bound to the membrane. The optimum time of UV irradiation is thus chosen accordingly (Fig. 9 & 12).

Covalently binding the genomic DNA fragments to the nylon membrane:

> Place the glass plate with Saran wrap and wet filter (face up, e. g. the side of the filter which was in contact with the gel during the transfer) under UV light and expose filter for the optimal time (see above).

> Put the membrane between two pieces of dry Whatman 17 paper and incubate in a preheated vacuum oven at $80-86°$ C for 10 minutes.

> If the filter is not to be used immediately, seal it between the Whatman papers under vacuum in a plastic bag (Melita) and store in the dark at $4°$ C.

Notes:

IMMOBILIZATION OF DNA ON NYLON MEMBRANES

Relationship between the Amount of Filter-Bound DNA and the Strength of the Hybridization Signal

As shown in Fig. 12 a maximal hybridization signal was obtained when roughly 30–50% of the DNA was stably bound to the filter membrane regardless of the hybridization buffer used. When 100% of the genomic DNA is bound to the filter by UV irradiation presumably a large proportion of the pyrimidines are covalently fixed onto the matrix. However, under these conditions the DNA is probably attached too rigidly to the filter and prevents optimal hybridization with the single-stranded labeled probe. We chose to use the Gene Screen membranes because they are very robust and gave the lowest background after hybridization. As seen in Fig. 12 different buffers were tested for the hybridization. Under our experiment conditions buffer A gave a much higher signal than buffer D for example. We decided, however, to use buffer D because it allows very high concentrations of the radioactive probe to be used (we tested up to 7×10^7 cpm/ml of hybridization mixture) without affecting the ratio between the background and the hybridization signals.

Hybridization of Immobilized DNA with Single-Stranded DNA Probe

The hybridizations are carried out with probes prepared as described in chapter III.12. A minimal hybridization volume is made possible by using the hybridization chamber illustrated in Fig. 13. It consists of two silicone-treated glass plates separated by a nylon fishing line, and held in place by metal clamps. Such a system is cheap and keeps the volume of the hybridization buffer as small as possible and permits the safe handling of the radioactive solutions. The filter does not have to be removed for prewetting, prehybridization and hybridization since all these steps are performed directly between the two glass plates. Once the hybridization is completed, the membrane is transferred to a plastic box. The hybridization buffer can be used twice if the total radioactivity is still high enough and the probe is not yet autoradiolyzed. In our experience it has not been possible to hybridize more than one filter at a time in this chamber, because the background of hybridization is increased to an unacceptable level (see Fig. 21/7 in 'Trouble-Shooting Guide').

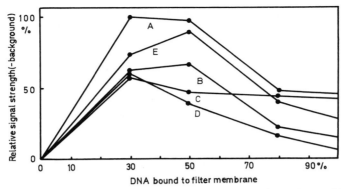

Fig. 12. Relationship between the amount of filter-bound DNA and the strength of the hybridization signal

Denatured chicken erythrocyte DNA fragments ($2\,\mu g/\mu l/0.2\,cm^2$ disc membrane) were spotted on Gene Screen^{TM} membranes and UV-irradiated for various times to give 30, 50, 80 and 100 % of stably bound DNA (UV 254 nm flux of $0.39\,mW/cm^2$ for 2.5, 5, 10, 45 min with a final heat treatment at 86° C for 30 min under vacuum). Nick-translated probe (Tsai et al., 1980; 10^5 cpm; specific activity: 6×10^8 cpm/μg) was used for hybridization in a total volume of $100\,\mu l$ covered with $70\,\mu l$ of paraffin oil in microtiter plates. Prehybridization and hybridibzation were as follows:

A Prehybridization was performed for 5 h at 42° C in $100\,\mu l$ of 4 × SSC, 50 % (v/v) formamide and 5 × Denhardt's solution, 0.02 M EDTA, 0.05 % (w/v) SDS, $500\,\mu g$ heparin/ml, 4 % dextran sulfate, $200\,\mu g$ of denatured *E. coli* DNA/ml. Hybridization with labeled DNA was carried out with the same buffer for 15 h at 42° C in a total volume of $100\,\mu l$.

B Prehybridization and hybridization were carried out as in (A) except that the 4 × SSC buffer was replaced by 0.25 M Na$_2$HPO$_4$ buffer (adjusted to pH 7.2 with H$_3$PO$_4$) and no sodium pyrophosphate was added.

C In the BioRad procedure (BioRad Instruction Manual), the filters were prewashed for 1 h in 0.2 × SSC, 0.5 % (w/v) SDS at 65° C. Prehybridization was done for 12 h 42° C in 5 × SSC, 0.05 M sodium phosphate (pH 6.5), 10 × Denhardt's solution, 1 mg denatured *E. coli* DNA/ml, 50° C (v/v) formamide. Hybridization was carried out for 25 h at 42° C in 5 × SSC, 0.025 M sodium phosphate (pH 6.5), 50 % (v/v) formamide, 2 × Denhardt's solution, $100\,\mu g$ of denatured *E. coli* DNA/ml.

D According to the procedure described by Church and Gilbert (1984), after rinsing with water, the filters were incubated for 5 min at 65° C in 0.25 M Na$_2$HPO$_4$ (adjusted to pH 7.2 with H$_3$PO$_4$), 0.001 M EDTA, 1 % (w/v) BSA (crystalline grade), 7 % (w/v) SDS. Hybridization was done for 24 h under the same conditions.

E According to the procedure described by Singh and Jones (1984), after rinsing with 4 × SET, the membranes were prehybridized for 3 h at 65° C in 4 × SET, 0.1 % (w/v) sodium pyrophosphate, 0.2 % (w/v) SDS, $500\,\mu g$ heparin/ml, 10 % (w/v) dextran sulfate. Hybridization was carried out for 15 h at 65° C in the same buffer supplemented with $100\,\mu g$ of denatured *E. coli* DNA/ml. (Saluz and Jost, 1986)

Reusing the Hybridization Probe

During hybridization usually only a small percentage of the radioactive probe is used for the hybrid formation. A good hybridization probe with a high specific activity can be used at least twice. Because of autoradiolysis and other degradative processes the second hybridization, however, should be performed without delay. The new membrane is prewetted and prehybridized as described in 'Step-by-Step Procedure' and then the preincubated radioactive buffer is poured straight into the chamber.

Control Sequence Used as Internal Standard for Hybridization

The control sequences obtained using the method described in chapter III.3 are used as an internal reference to test the quality of the hybridization. The use of sufficient control DNA gives slightly higher signals than are obtained with the genomic DNA so that after overnight autoradiography of the hybridized filter, it is possible to get preliminary information about the quality of the experiment and the filter membranes. The best exposure time for all other sequencing lanes can be estimated after such a short exposure.

Materials and Buffers

> Hybridization buffer: 12.5 ml of 0.5 M Na_2HPO_4, pH 7.2; 50 μl of 0.5 M EDTA, pH 8.5; 1.75 g of sodium dodecylsulfate; 0.25 g BSA (cristalline grade); H_2O added to 25 ml

> Silicone-treated glass plate ($180 \times 400 \times 4$ mm)

> Silicone-treated glass plate ($180 \times 395 \times 4$ mm)

> Nylon fishing line (diameter: 0.35 mm)

> Silicone grease (Bayer, medium viscosity)

> Metal clamps

> Plastic box (approx. $300 \times 350 \times 50$ mm)

> Saran wrap

> Conical falcon tube (15 ml)

> Water bath (90° C)

> Oven for hybridization (for temperature requirement see equation VIII, p. 76)

> Pasteur pipettes

Step-by-Step Procedure

> Prepare 25 ml of the hybridization buffer as
  follows:

  – Add to 12.5 ml stock solution (0.5 M
    Na_2HPO_4, titrated with ortho-phosphoric acid
    to pH 7.2)
  – 50 μl of 0.5 M EDTA, pH 8.5
  – 1.75 g sodium dodecylsulfate (NaSDS; 7%)
  – 0.25 g BSA (1%; crystalline grade) (add BSA
    and NaSDS (in aliquots), while stirring with a
    magnetic stirrer).
  – Make up to 25 ml with water and stir well at
    room temperature until all the ingredients are
    completely dissolved.
  – Prepare 2 silicone-treated glass plates 18 × 40
    cm using a solution of 1% dimethyldichloro-
    silane in carbontetrachloride; plates have to be
    spotless.
  – Assemble the hybridization chamber as out-
    lined in Fig. 13:

(1) Apply a thin layer of silicone grease (Bayer,
    medium viscosity) to the longer glass plate
    with a syringe.

(2) Take a nylon fishing line (normal diameter is
    0.35 mm but a thinner one can be used for spe-
    cial purposes). Tape the end of the thread to
    the bench at the top of the plate and pull the
    thread onto the silicone grease. Guide the
    thread onto the plate using a plastic pipette tip.
    The fishing line can be replaced by silastic
    tubing used in surgery. In this case no silicone
    grease is needed.

(3) Place the dry membrane on the glass plate.

(4) Cover with the second glass plate (about 0.5 cm shorter than the lower one) and fix in place with metal clamps as indicated in figure. The metal clamps should hold the plates just on top of or inside but not on the outer side of the nylon thread.

(5) Rinse the nylon membrane with water using a 10 ml pipette, pour out water (the residual water will not interfere) and replace it with hybridization buffer lacking the labeled probe. Put the hybridization chamber in a slanted position, so that the buffer cannot run out, into a prewarmed plastic box and incubate for about 5 minutes at the hybridization temperature in an oven (58° C). Replace the buffer with

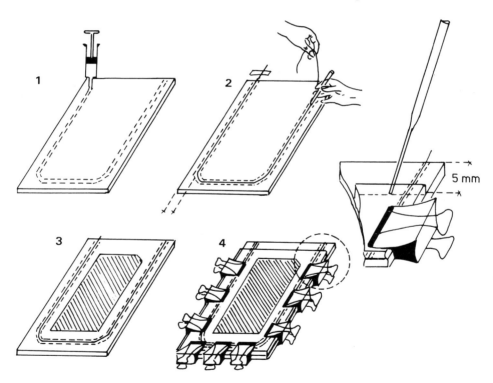

Fig. 13. Assembling the hybridization chamber

The hybridization system is assembled as described in 'Step-by-Step Procedure'.

PREHYBRIDIZATION AND HYBRIDIZATION

the hybridization buffer containing labeled probe (for our experimental conditions about 10 ml). Seal the chamber with Saran wrap and put it back, still in a slanted position, into the box and then incubate for 18–24 hours at 58°C. The hybridization buffer containing the probe was prepared by adding the probe, pre-heated to 90°C for 1–2 minutes in an Eppendorf tube, to 10 ml of hybridization buffer in a conical Falcon tube (15 ml). Vortex and use immediately for hybridization.

Further reading on the practical and theoretical aspects of hybridization:

> Nucleic Acid Hybridization, edited by B.D. Hames and S.J. Higgins (1985), IRL PRESS, Oxford–Washington DC (and references therein).

> Biophysical Chemistry, edited by Cantor and Schimmel (1980), W.H. Freeman and Co., U.S.A. (and references therein).

Notes:

Notes:

8 Processing of the Hybridized Filters

The Washing Procedure

After hybridization the filters are washed several times under stringent conditions to remove the excess labeled probe. As explained in the last chapter on hybridization, the stringency of the buffer can easily be increased by
lowering the molarity of the monovalent cations and/or increasing the temperature. For all the genomic sequencing experiments carried out in our
laboratory, the washing buffers described in the 'Step-by-Step Procedure'
were sufficient to give a low background. The molarity of sodium ion is kept
at 74 mM for the more stringent washing buffer (washing buffer 2). The
optimal temperature was determined empirically in the same way as
described in the chapter III.7. The melting temperature (Tm) can be
obtained starting with the formula (VIII, page 72) for hybridization and
74 mM sodium ion concentration:

(IX) $\quad Tm = 16.6 \times log\ 0.074 + 0.41\ (\%G+C) + 81.5 - 10$

which can be simplified to

(X) $\quad Tm = 52.7 + 0.41\ (\%G+C)$

As the optimal washing temperature (WT) was experimentally found to be
19.7° C below the Tm a general practical formula for the determination of
the washing temperature (used for washing buffer 1 and 2) of the genomic-
sequencing membranes can be determined:

(XI) $\quad WT = 33° C + 0.41\ (\%G+C)$

Note: *We have observed that a* WT *of 48° C was sufficient for most experiments (all the sequences tested with this* WT *had a (G+C)-content of
30–47%).*

The best way to wash the filter is to incubate the buffers in glass bottles in a
waterbath at the calculated washing temperature (WT). The volumes of the
washing buffers should be kept as small as possible, e. g. the membrane in
the plastic box should just be covered. The washing itself is performed at
room temperature on a slow moving shaker. Under these conditions the buf-

fer cools down very quickly. At room temperature, in a closed polystyrene box (described in 'Materials and Buffers'), the temperature of 100 ml buffer will drop within 5 min from 47° C to 33° C. This procedure is relatively mild since the stringency is decreased by lowering of the temperature of the buffer. The first two washes with buffer 1 are carried out for 5 minutes each. Buffer 2 is used until no background can be detected on specified regions of the membrane. Usually 5 to 7 washes of 5 minutes each are sufficient. When hybridization signals are lost or when the background remains too high, the stringency of the washing procedure should be either changed by altering either the buffer ionic strength or the temperature of the washing.

In successful genomic sequencing experiments the amount of radioactivity on the filter decreased very rapidly during the first washes and then more slowly towards the end of the washing procedure.

Reuse of the Membranes

Rehybridization of the membranes with a probe complementary DNA strand often yields useful information. However, before a filter can be reused for a new hybridization it is absolutely necessary to strip off the old probe from the filter. The buffer described in the 'Step-by-Step Procedure' is usually sufficient to remove the old probe. In extreme cases the stringency can be augmented by increasing the temperature and/or by using distilled water. Sometimes, especially when DNA is covalently bound to filters, problems in removing the old probe can arise (Cannon et al., 1985). In this case it is advisable to wait for the ^{32}P-label to decay for several half-lives until no radioactivity can be measured on the filter. Only then are the membranes washed as indicated in the 'Step-by-Step Procedure'. To avoid such situations it is important that the filter is not allowed to dry completely after the final wash. If semi quantitative results are needed, it is advisable to repeat the genomic sequencing several times in duplicates. Two aliquots from the chemical cleavage reactions are separated on two different gels, transferred and hybridized: one filter with the probe for the upper DNA strand and the other with the probe for the lower DNA strand.

Materials and Buffers

> Water bath

> Shaker

> 1 polystyrene box ($330 \times 220 \times 50$ mm)

> Saran wrap

> Funnel

> Erlenmeyer flask

> Whatman 17 filter paper

> Stock buffer (0.5 M Na_2HPO_4, titrated with ortho-phosphoric acid to pH 7.2)

> Washing buffer 1 (20 mM Na_2HPO_4, pH 7.2; 1 mM EDTA; 5 % sodium dodecylsulfate; 0.5 % BSA fraction V)

> Washing buffer 2 (20 mM Na_2HPO_4, pH 7.2; 1 mM EDTA; 1 % sodium dodecylsulfate)

Step-by-Step Procedure

> Hybridized filters are washed with the following solutions:

Buffer 1:

- To 20-ml hybridization stock buffer (0.5 M Na_2HPO_4, titrated with ortho-phosphoric acid to pH 7.2) add 1 ml of 0.5 M EDTA, pH 8.5, 25 g sodium dodecylsulfate, 2.5 g bovine serum albumin (fraction V), make up to 500 ml with water.
- Stir well until everything is dissolved.

Buffer 2:

- Add 2 ml 0.5 M sodium EDTA, pH 8.5, 10 g sodium dodecylsulfate to 40 ml hybridization stock buffer, dissolve and add water to 1000 ml.
- Keep the washing buffers 1 and 2 for at least 20 minutes in a water bath at the calculated temperature (see formula XI).
- Remove the hybridization chamber from the oven, take off the Saran wrap and clamps and pour out the radioactive buffer into a sterile Erlenmeyer flask by using a funnel (the buffer can be used immediately for a second hybridization).

> Open the hybridization chamber, take the filter out (use forceps) and put it into the plastic box as (described in 'Materials and Buffers') containing about 100 ml of the prewarmed washing buffer 1. Make sure that the filter is completely covered with buffer.

> Put the box on a shaker (not more than 40 rpm) at room temperature for 5 minutes. Replace buffer 1 with ca. 100 ml of the same buffer and shake again 5 minutes.

> Replace buffer 1 with an equal volume of pre-warmed washing buffer 2 and shake (40 rpm) for 5 minutes at room temperature.

> Repeat the last step 5–7 times or until no radioactivity can be detected with a hand monitor on an area of the nylon membrane which contains no bound DNA.

> Lay the filter on a clean Whatman 17 paper until the excess buffer is absorbed. The nylon membrane should not be allowed to dry completely. The filter is placed face down on a piece of Saran wrap, and a fresh piece of Whatman 17 paper (just slightly bigger than the filter) is placed on top of the filter and the ends of the Saran wrap folded over. The filter is now ready for X-ray film exposure. The final decision as to whether the background is low enough can be made after short X-ray film exposure (24 h) to the filter.

Notes:

9 Autoradiography and Photography

Autoradiography

From the preceding chapters it is clear that the ultimate aim of the genomic sequencing procedure is to obtain strong hybridization signals and a high resolution of the sequence of interest. Considering its overall complexity and the small amount of genomic target DNA, each optimized step will contribute to the quality of the final result. The choice and treatment of the X-ray films also play a crucial role. A signal obtained after several days exposure of the X-ray film, may tempt the inexperienced genomic sequencer to use a more sensitive film, for example the XAR-5 from Kodak, without considering that the large size and lower density of the silver-halide crystals in such a film will drastically reduce the resolution of the autoradiogram. If we take the other extreme, a film with a very high density of small silver-halide grains (such as Industrex-M from Kodak) would give an ideal resolution, but because of its lower sensitivity it would not be adequate for genomic sequencing. For these reasons we recommend the use of X-ray films of middle sensitivity that still give very good resolution (X-OMAT S from Kodak or equivalent). It is interesting to note that the Kodak direct exposure film DEF-5 used for X-ray diffraction studies is 2 to 3 times faster than the X-OMAT AR film or 4−6 times more sensitive that the X-OMAT S film from Kodak. Due to its very thin layer and high density of silver-halide grains the DEF film has a power of resolution comparable to the X-OMAT S film. However, the DEF films are only available in sheets of up to 19×24 cm and are more expensive than the X-OMAT S films.

The sensitivity of the films can be increased by preflashing the film and using an intensifying screen (except for DEF-films) during the exposure. Even though the intensifying screen has the tendency to lower the resolution of ^{32}P-labeled molecules it should still be sufficient for genomic sequencing. When using intensifying screens it is important to keep a very close contact between the filter and the film and to use the correct temperature. Otherwise both the resolution and the sensitivity are bound to decrease. According to Laskey (1984) the optimal temperature range for exposing an X-ray film is between -40 and $-90°$ C with an optimum at $-78°$ C. If a preflashed (A_{540nm}

= 0.2) Kodak X-OMAT R^{RM} film is used, the improvement of the sensitivity for ^{32}P is approximately tenfold. Preflashing film to an absorption (540 nm) of 0.1–0.2 and lowering the temperature to (−40)–(−70° C) gives the film a linear response to the emitted light which is very important for semi-quantification of the results. Semi-quantitation of an X-ray film based on the use of a non-preflashed film may give very large mistakes for very low radioactive intensities since under these conditions the absorbance of the film image is a sigmoid function and not a linear one (see Fig. 14).

Intensifying screens work by converting the beta emission from the isotope to light energy. However they can also absorb energy from room lights and emit this energy as light. This 'after-glow' effect may fog the film and is best avoided by keeping the screen in the dark for several hours before starting an autoradiographic exposure.

The sensitivity of an X-ray film can be further enhanced (factor 2−4) by baking the film at 65° C in an 8 % 'forming-gas' environment (8 % H$_2$ + 92 % N$_2$) as suggested by Philips et al. (1986) and Smith et al. (1985). Since this treatment and preflashing act on the film in different ways, their effects are additive.

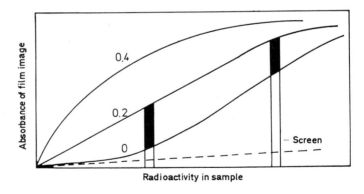

Fig. 14. Pre-flashing of X-ray films

Requirement for pre-exposure to obtain a linear response of film to light from ^{32}P with an intensifying screen. The effect of pre-exposure absorbances of 0, 0.2 and 0.4 above the absorbance (540nm) of unflashed film are shown. The vertical bars illustrate how the effect of pre-exposure (black area) is easily underestimated when it is assessed only with large amounts of radioactivity. Reprinted courtesy of R. A. Laskey (1984).

Photography of X-ray Films

Even the best genomic-sequencing results can be ruined by poor photography. In this section we give some hints about the photography of X-ray films.

The best results we obtained with high-quality, panchromatic negative films which have an increased sensitivity in the red end of the spectrum, high resolution and extremely fine silver-halide crystals. All these parameters are fulfilled by a negative film from Kodak, the Kodak Technical Pan 2415/6415. As the contrast of this black-and-white film can be altered by changing the developing conditions, it is extremely well suited for copying X-ray films. Most of the copies in this book were made using a Nikon camera (F3) fitted with an original 55 mm macro lens (diaphragm: 16; with and without a yellow Kodak-Wratten filter No 16) and Kodak Technical Pan 2415-negative film, was exposed at 50 ASA (automatic; variants $+\frac{1}{3}$, 0, $-\frac{1}{3}$) and developed in Kodak Technidol (powder, Cat No 1010628) as described in the 'Step-by-Step Procedure'. The positive prints were made with Ilfobrom paper (3–4) which we developed in Neutol from Agfa Gaevert. The X-ray film to be photographed is placed on a photographic light table with an opal glass screen (common ground glass screens have too high a grain). Because of diffused light, the areas around the sequencing tracks have to be covered with black opaque cardboard and during exposure all extraneous lights have to be switched off.

Materials

Autoradiography

Preparation of the electronic flash unit:

> Commercial electronic flash

> Kodak-Wratten filter No. 21 or 22

> Whatman 1 filter paper

> Adhesive tape

Preflashing of the X-ray films:

> X-ray film

> Prepared flash unit

> Developing machine or containers for manual developing

> Photometer (540 nm)

> Cuvettes

> Scissors

Autoradiography of the processed filter membrane:

> Prepared flash unit

> Film cassette

> Intensifying screen (Kodak X-Omatic Regular or any equivalent)

> X-ray film

> Freezer (-40 to $-70°$ C)

> Developing machine or containers for manual developing

Photography

Developing of the negatives:

> Developer tank

> Kodak-Technidol powder

> Thermometer

> Stop watch

> Kodak Fixing Bath F-5

> Drying-room

Developing of the positive prints:

> Ilfobrom paper (grade 3 or 4)

> Neutol

> Fixing bath

> Drying machine

Step-by-Step Procedure

Autoradiography

Preparation of the electronic flash:

> Take a commercial electronic flash unit. The duration of flash should not be longer than 1 msec; longer flash times result in a decreased background-sensitivity ratio (Laskey and Mills, 1977).

> Cut a Kodak-Wratten filter No. 22 or 21 to the precise size of the flash window. Cut 1–2 sheets of Whatman 1 filter paper (diffuser) to a slightly larger size than the flash window.

> Put the Kodak-Wratten filter onto the flash window and then cover this with the Whatman papers and then fix the edges of the paper to the flash unit with adhesive tape, as indicated in Fig. 15. The unit is ready for use.

Fig. 15. Electronic flash unit

A commercial flash unit prepared as described by Laskey (1984). The Kodak-Wratten filter and the Whatman 1 papers are fixed on the flash window by means of an adhesive tape.

Preflashing X-ray films:

> A test film is fixed to a wall in the darkroom (the surface of the wall should be smooth; otherwise put a paper behind the film). Determine the optimal light intensity by exposing strips of the film to a successively increasing number of flashes from a fixed distance (appr. 70 cm) and by measuring the optical density (540 nm) of the developed film. The film strip with an optical density of 0.1–0.2 units at 540 nm gives the correct number of flashes to be used (at least for films from the same lot).

Autoradiography of the processed genomic sequencing filter membrane:

> Preflash an X-ray film as described above.

> Put the preflashed side of the X-ray film directly on the intensifying screen in a film cassette.

> Put the upper side of the filter membrane (filter prepared as described in chapter III.8) onto the X-ray film and expose at $-70°$ C for approximately 1 day.

> In the darkroom take out the film and fix it on a frame for immediate manual development. If the film has to be developed by an automatic machine, hold the film for a couple of minutes to equilibrate at room temperature (avoid exposure to darkroom safelights) and then place it in the automatic developing machine.

> This film is now used to decide the correct exposure time of the filter. Our shortest exposure time was 4 hours for the control sequences and 12 hours for the genomic sequences. Experiments of medium signal strength need between 1 and 2 days exposure for the control sequence and 4–10 days for the genomic sequences (preflashed X-OMAT S).

Photography

Developing of Kodak technical pan film 2415 (half-tone):

> Prepare the Technidol developer a short time before starting the developing process (developer in solution is unstable): Dissolve 15 g of the Technidol powder in 415 ml of warm water (50° C). Adjust final volume to 472 ml with warm water (50° C) and allow to mix well. Cool solution to 20° C (tolerance for shifts in temperature is small).

> Load the film into a developing tank (total darkness is required). Fill the tank with the developer solution. Dislodge any air bubble by tapping the bottom of the tank on the work bench surface. Shake the tank immediately for 15 seconds. Put the tank on the table for 28 seconds and then agitate it for 2 seconds. Repeat the second agitation every 30 seconds for 15 minutes.

> Discard the developer solution and immediately fill the tank with water and shake the tank vigorously for 30 seconds.

> Discard the water and fill the tank with Kodak Fixing Bath F-5 (20° C). Agitate tank frequently during 4 minutes.

> Discard fixing bath and rinse film in open box with clear running water (19−21° C) for 15−20 minutes. Replace water with distilled water for 2−5 minutes and dry the film in a dust-free place at 49−60° C.

Conditions for positive prints:

> Use a grade 3 or 4 paper (Ilfobrom; Ciba Geigy) for exposure.

> Develop exposed papers in Neutol for 2 min (dilution 1 : 10 or 3 minutes in a 1 : 20).

> Stop in running water.

> Fix for 5−10 minutes in a Kodak fixer.

> Place photos in running water for a least 30 minutes.

> Dry pictures in a drying machine.

Notes:

10 Cloning of DNA Probe in M13

This cloning procedure has already been described by Messing et al. (1983). The purpose of this chapter is to help those who do not routinely use this technique.

Materials and Buffers

> M13-vectors cut with suitable restriction enzyme(s)

> Cloned experimental DNA

> *E. coli* JM 101 strain or equivalent

> Solutions for approx. 10 minimal medium agar plates:
>   1. M9 salts: 1.2 g Na_2HPO_4; 0.6 g KH_2PO_4; 0.1 g NaCl; 0.2 g NH_4Cl; add H_2O to 100 ml; autoclave
>   2. 0.2 ml 1 M $MgSO_4$; 0.2 ml 0.1 M $CaCl_2$; 0.2 ml 1 M thiamine-HCl, sterile filter
>   3. 100 ml 3 % bactoagar

> T_4-DNA ligase (blunt ends: 10 units/μl; cohesive ends: 3 units/μl) 10 × ligase buffer (500 mM TrisHCl, pH 7.8; 100 mM $MgCl_2$; 200 mM dithiothreitol; 10 m ATP (500 μg BSA/ml incubation mixture).

> 2 × YT-medium: 16 g tryptone (bacteriological grade), 10 g yeast extract; 5 g NaCl; H_2O to 1000 ml

> Incubator (37° C)

> Spectrophotometer

> Agar (bacteriological grade)

> Agarose (Type I)

> 5-Bromo-4-chloro-3-indoxyl-beta-D-galactoside (Xgal)

> Dimethylformamide (analytical grade)

> Isopropyl-beta-D-thiogalactopyranoside (IPTG)

> Sterile toothpicks (bacteriocide-free) or sterile plastic pipette tips

> Eppendorf centrifuge

> Sorvall centrifuge

> HB4-rotor or equivalent

- \> Competence solutions A and B:
  A: 10 mM NaCl; 50 mM $MnCl_2$; 10 mM sodium acetate, pH 5.6
  B: 75 mM $CaCl_2$; 100 mM $MnCl_2$; 10 mM sodium acetate, pH 5.6

- \> Polyethylenglycol 6000 (PEG 6000; Serva)

- \> Tris-EDTA-buffer (TE): 10 mM Tris-HCl, pH 7.5; 1 mM EDTA

- \> Phenol saturated with 1 M Tris, pH 8

- \> Ethanol

Step-by-Step Procedure

Ligation of double-stranded probe molecules with the M13 RF DNA:

> Prepare 10 × ligation buffer: 500 mM Tris-HCl, pH 7.8; 100 mM MgCl$_2$; 200 mM dithiothreitol; 10 mM ATP (500 μg BSA/ml incubation mixture). Mix 2 μl of this 10 × buffer, 1 μl of cut M13 RF DNA (0.8 pmoles/ml), 1 μl probe DNA (4 pmoles/ml) and 15 μl of sterile distilled water in an Eppendorf tube. Mix by tapping the tube and spin reaction tube for few seconds in an Eppendorf centrifuge. Add 1 μl of T$_4$ ligase (3 units/μl for cohesive ends; 10 units/μl for blunt ends) mix by tapping the tube, spinning for a few seconds (Eppendorf centrifuge) and incubate at 16° C for 2 hours for cohesive ends and 12 to 24 hours for blunt ends. The ligated DNA is kept frozen until used.

Preparation of competent *E. coli* cells (JM 101):

> Prepare minimal medium agar plates:
* M 9 salt: 1.2 g of Na$_2$HPO$_4$; 0.6 g of KH$_2$PO$_4$; 0.1 g of NaCl and 0.2 g of NH$_4$Cl are dissolved in a total volume of 100 ml. Autoclave M9 salts. Add 0.2 ml of 1 M MgSO$_4$; 0.2 ml of 0.1 M CaCl$_2$; 0.2 ml of 1 M thiamine-HCl (sterile filter) and 1 ml 40 % glucose.
* Three g of bactoagar are autoclaved in 100 ml of water and mixed at 50° C with the solution described above. Two hundred ml of minimal medium is sufficient for 10 petri dishes (diameter: 8.5 cm).

> Grow JM 101 on minimal medium plates overnight at $37°$ C (colonies are viable for several weeks at $4°$ C).

> Prepare $2 \times$ YT medium: 16 g/l casein hydrolysate (tryptone), 10 g/l yeast extract and 5 g/l of sodium chloride, use tap water and adjust the pH to 7.2–7.4, if necessary with sodium hydroxyde and sterilize by autoclaving.

> Inoculate 3 ml of $2 \times$ YT medium with a single colony from the minimal medium plate and incubate at $37°$ C overnight. Add $330 \mu l$ of this culture to 50 ml of $2 \times$ YT and incubate in $37°$ C shaking incubator until the OD_{550} is between 0.2–0.3 (this should be sufficient for 10 transformation reactions). Chill the cells on ice and harvest them at 8000 x g at $4°$ C for 10 minutes.

> Resuspend the cells in 10 ml of prechilled ($4°$ C) competence solution A: 10 mM sodium chloride/ 50 mM manganese chloride/10 mM sodium acetate, pH 5.6 (the solution containing $MnCl_2$ cannot be autoclaved; use sterile filtration only). The cells have to be chilled on ice for 20 min and harvested as before. The pellet is resuspended in 1 ml of sterile ice-cold 'competence' solution B: 75 mM calcium chloride/100 mM manganese chloride/ 10 mM sodium acetate, pH 5.6. The cells should be either used within 60 min of being resuspended or kept for a few weeks in the competence solution B + 5 % glycerol at $-70°$ C.

Preparation of plating cultures:

> Inoculate 3 ml of $2 \times$ YT-medium with a single JM 101 colony from a minimal medium agar plate at $37°$ C with shaking until OD_{550} is approximately 1. Place the culture on ice. A 3-ml culture is sufficient for 15 transformation plates.

Transformation:

> Add 3 μl of ligase-reaction mixture (10 ng of the vector) to 100 μl of competent cells (ice-cold) and incubate on ice for 30 min by mixing from time to time. Incubate for 2 min at 37° C and plate out immediately.

Plating procedure:

> Prepare 3 sterile tubes (13 × 100 mm) for each transformant. Add 200 μl of plating culture to each tube. Add 1 μl of transformation reaction to the first tube, 10 μl to the second and 90 μl to the third. To each tube add 3 ml of melted 0.7 % agarose (agarose gives better plaques than agar) maintained at 45° C and containing 100 μl of 2 % Xgal in dimethylformamide and 20 μl of 100 mM IPTG; the latter two substances are only added when the top agarose has cooled to 45–55° C. Pour the warm agarose solution (kept in a water bath) onto a YT-plate (15 g agar/liter YT medium; 1 liter is sufficient for 50 plates). Leave the plates for about 1 h at room temperature and then incubate overnight at 37° C.

Preparation of the template DNA (micro-scale preparation):

> Touch the middle of a well-separated colorless plaque with a sterile toothpick or micropipette tip and inoculate 2 ml of 2 × YT-medium containing 10 μl of freshly prepared JM 101 plating culture. After not more than 5–8 hours incubation in a shaker at 37° C, 1.5 ml of the culture is transferred into an Eppendorf tube and centrifuged for 5 min in a microcentrifuge. 1 ml of the supernatant containing the phage is transferred into another Eppendorf tube. The rest of the supernatant is frozen at −20° C and can be used over several weeks for infecting further cultures such as those used in

the large-scale preparations described in the next chapter. The pellet can be used for the isolation of the RF-DNA if required. Add $250\,\mu$l of 20% polyethylenglycol/2.5 M NaCl (PEG: w/v; PEG 6000) to 1 ml of the supernatant and mix gently by inverting the tube. Leave the tube at room temperature for 15 min (longer periods allow the precipitation of other small unwanted molecules). Centrifuge the tube for 10 min in a microfuge, remove supernatant carefully and centrifuge for another 30 seconds to separate the pellet from the rest of PEG 6000. Dissolve the pellet in $100\,\mu$l TE buffer (10 mM Tris-HCl, pH 7.5; 1 mM EDTA) by tapping the tube and extract twice with phenol ($50\,\mu$l of saturated phenol with 1 M Tris, pH 8). The aqueous phase is transferred to another Eppendorf tube and mixed with $9\,\mu$l of 3 M sodium acetate and 3 volumes of ethanol ($270\,\mu$l). After incubating for 15 min in dry ice/ethanol centrifuge the tube in a microfuge for 10 minutes. The pellet is washed twice with 70% ethanol (no salt added). After drying the pellet in a speed vac the single-stranded DNA is redissolved in either water or $1\times$ TE (10 mM Tris-HCl, pH 7.5/1 mM EDTA). After measurement of the OD_{260} the DNA can be used for the synthesis of the hybridization probe.

Notes:

11 Large-Scale Preparation of Cloned DNA in M13

This procedure was originally described by Messing (1983). One large-scale preparation yields enough single-stranded DNA for many hybridization experiments.

Materials and Buffers

> Freshly prepared *E. coli* strain JM 101 (grown on a minimal medium plate)

> Supernatant containing the single-stranded form of M13 from a micro-scale experiment

> Shaker (37° C)

> Sorvall centrifuge or equivalent

> GSA-rotor (Sorvall) or equivalent

> SS-34-rotor (Sorvall) or equivalent

> 5 M sodium chloride

> 50 % polyethylenglycol 6000 (PEG 6000) in H_2O

> 1 × Tris-EDTA-Sarcosyl (10 mM Tris-HCl, pH 7.5; 1 mM EDTA; 0.5 % sarcosyl)

> 1 × Tris-EDTA (TE; 10 mM Tris-HCl; pH 7.5; 1 mM EDTA)

> Phenol saturated with 1 M Tris, pH 8

> 3 M sodium acetate (NaOAc), pH 5; 0.005 M EDTA

> Chloroform

> Ethanol

Step-by-Step Procedure

Large-Scale Preparation of M13 (Calculated for One-Liter Culture)

> Prepare 10 ml of an *E. coli*-JM 101 culture (OD_{550} greater than 1) by inoculating fresh cells from a minimal medium plate.

> Place 10 ml of this starter culture into 1 liter of steril 2 × YT medium (16 g/l of casein hydrolysate (tryptone), 10 g/l of yeast extract and 5 g/l of sodium chloride and adjust to pH 7.2–7.4 if necessary with sodium hydroxide. The flask should contain less than $^3/_{10}$ its volume of culture medium to ensure good aeration.

> Shake the culture at 150 rpm for approx. 4 hours at 37° C until OD_{550} reaches 0.45.

> Infect the culture with 1 ml of M13 containing supernatant from a micro-scale preparation as described in the previous chapter.

> Incubate in a shaker (150 rpm) for approx. 4 hours and centrifuge the culture at 4° C for 10 minutes at 10 000 × g in a GSA-rotor (Sorvall).

> Discard the pellet and store the supernatant at 4° C overnight.

> **Note:** *The pellets can be used for the isolation of the RF-form (replicative form of the recombinant DNA) if required.*

> Aliquots of the supernatant can be kept frozen (− 70° C) and used for further infections.

> Put supernatant fraction on ice for approx. 30 minutes and mix it gently with 100 ml of 5 M sodium chloride (NaCl) added slowly in 10–15 ml aliquots. Similarly add 100 ml of 50% polyethylenglycol (PEG 6000) in 10–15 ml aliquots.

> Stir gently for 2 minutes at room temperature.

> Leave the suspension on ice for 2 hours.

> Centrifuge at 4°C at 10 000 × g (GSA-rotor, Sorvall) for 15 minutes.

> Decant off the supernatant (which can be used for phage titer determination) and resuspend the sediment gently overnight in 40 ml of TE/Sarcosyl buffer (10 mM TrisHCl, pH 7.5; 1 mM EDTA; 0.5% Sarcosyl) at 4°C.

> Place the phage suspensions on ice for 30 minutes.

> Slowly add 4 ml of 5 M NaCl and 4 ml of 50% polyethylenglycol (PEG 6000) to each phage suspension.

> Stir gently for 2 minutes at room temperature.

> Leave the suspensions on ice for 2 hours.

> Centrifuge for 30 minutes at 4°C at 12 000 × g (SS-34 rotor; Sorvall).

> Gently resuspend each sediment in 4 ml of TE (10 mM Tris-HCl, pH 7.5; 0.1 mM EDTA) overnight at 4°C on a rocking platform.

> Add 2 ml phenol (saturated with 1 M Tris, pH 8) and add 2 ml of chloroform to the suspension, mix well by inverting the tube.

> Centrifuge for 5 minutes at 12 000 × g in a SS-34 Sorvall rotor.

> Repeat the phenol chloroform extraction twice.

> Add ¹⁄₁₀ volume of 3 M sodium acetate, pH 5; 0.005 M EDTA and 2.5 volume ethanol and mix gently by inversion and leave it at $-20°$ C overnight.

> Centrifuge DNA in a Beckman SW-27 rotor for 2 hours at 25 000 rpm at 4° C.

> Decant the supernatant and dry the pellets under vacuum.

> Resuspend the pellets very gently in $300\,\mu$l TE (10 mM Tris-HCl, pH 7.5/1 mM EDTA).

> Determine OD_{260}.

> Add TE to give a final concentration of $6\,\mu$g DNA/μl TE.

> Store in $15\,\mu$l aliquots at $-70°$ C.

12 Synthesis of Oligonucleotide Primers and Single-Stranded Labeled Probes

The synthetic primer required for the synthesis of the labeled single-stranded DNA probe can easily be synthesized in automatic synthesizers or by hand (Sproat and Gait, 1984). Otherwise it is possible to buy such DNA primers from some of the molecular biological companies. An advantage of having a large piece of DNA cloned in M13 is that different synthetic primers can be used to produce different specific single-stranded DNA probes. Commercially available primers are usually complementary to the 3′ flanking region of the M13 polylinker. New England Biolabs, for example, sells a 17-Mer oligonucleotide 5′d(*GTAAAACGACGGCCAGT*)3′ that can easily be used as a primer for the synthesis of the labeled DNA probe.

Since the specific activity and purity of the labeled DNA probe has to be as high as possible, it is not satisfactory to use any form of end-labeled probe. Therefore the pulse chase technique is used to generate a primed-extended DNA strand. A strong pulse of (^{32}P-alpha)dATP is given during the first few minutes of incubation. This ensures the incorporation of the labeled dATP into the probe. The primed-extension is then completed by chasing the reaction with a vast excess of unlabeled dATP. Since the total concentration of dATP will be above the Km of the reaction, this will ensure the efficient synthesis of an homogeneous population of DNA molecules, that can be cleaved with a suitable restriction enzyme. The presence of an M13 'tail' at the 3′-end of the probe increases the strength of the hybridization-signal and is therefore important. One useful restriction site is the *Pvu*II which creates an M13 'tail' approximately 180 nucleotides long. Fig. 16 shows a restriction map of this area in M13mp18. The restriction endonuclease *Pvu*II can always be used providing that there is no corresponding recognition site for this enzyme within the probe. Although the 'tail' can be very useful for increasing the strength of the hybridization signal, it should not be used when there is a strong homology with other sequences of the genome to be studied. The probability that such a homology occurs is however very small. A computer-homology search with this stretch of M13-DNA through 10^6 sequence residues of the EMBL library revealed that there are only very few known homologies in eucaryotic genomes. The synthesized probe has to be purified prior to use as described in the next chapter. Fig. 17 shows an

```
        1  26           H      1  6 3 11              1
                        /             /   /           /
        GGCAAACCAGCGTGGACCGCTTGCTGCAACTCTCTCAGGGCCAGGCGGTGAAGGGCAATC
5900    +---------+---------+---------+---------+---------+---------+    5959
        CCGTTTGGTCGCACCTGGCGAACGACGTTGAGAGAGTCCCGGTCCGCCACTTCCCGTTAG


                                              BS
        AP                         H      scBANHHN
        lv                         p      trahahal
        uu                         h      NFnaraea
        12                         1      11121124
         /                         /       / /
        AGCTGTTGCCCGTCTCGCTGGTGAAAAGAAAAACCACCCTGGCGCCCAATACGCAAACCG
5960    +---------+---------+---------+---------+---------+---------+    6019
        TCGACAACGGGCAGAGCGACCACTTTTCTTTTTGGTGGGACCGCGGGTTATGCGTTTGGC


                        H                F
        TMHT    C H     i                n AP         B
        hnhh    f a     n                u lv         b
        alaa    r e     f                4 uu         v
        1111    1 3     1                H 12         1
           /                            /
        CCTCTCCCCGCGCGTTGGCCGATTCATTAATGCAGCTGGCACGACAGGTTTCCCGACTGG
6020    +---------+---------+---------+---------+---------+---------+    6079
        GGAGAGGGGCGCGCAACCGGCTAAGTAATTACGTCGACCGTGCTGTCCAAAGGGCTGACC


                                                              BS
                  H                      A          B N       sc
                  h                      l          a l       tr
                  a                      u          n a       NF
                  1                      1          1 4       11
                                                              /
        AAAGCGGGCAGTGAGCGCAACGCAATTAATGTGAGTTAGCTCACTCATTAGGCACCCCAG
6080    +---------+---------+---------+---------+---------+---------+    6139
        TTTCGCCCGTCACTCGCGTTGCGTTAATTACACTCAATCGAGTGAGTAATCCGTGGGGTC


                  H
                  p
                  a
                  2
        GCTTTACACTTTATGCTTCCGGCTCGTATGTTGTGTGGAATTGTGAGCGGATAACAATTT
6140    +---------+---------+---------+---------+---------+---------+    6199
        CGAAATGTGAAATACGAAGGCCGAGCATACAACACACCTTAACACTCGCCTATTGTTAAA


                                  E
                  A           N   c   T  A BsBgNRAHKNNaSNaccSMX
                  l           l   o   a  l apailsvppccmalurrmab
                  u           a   R   q  u n1nAaaaaniiHca3FFaea
                  1           3   1   1  1 22114112111114A11111
                                         / /  /  / /// / //// /
        CACACAGGAAACAGCTATGACCATGATTACGAATTCGAGCTCGGTACCCGGGGATCCTCT
6200    +---------+---------+---------+---------+---------+---------+    6259
        GTGTGTCCTTTGTCGATACTGGTACTAATGCTTAAGCTCGAGCCATGGGCCCCTAGGAGA


        H  H              B   H
        Xi SAiMT          P   s  NSi A          C H
        hn acnna          s   p  lpn l          f a
        of lcclq          t   M  ahd u          r e
        21 11211          1   1  313 1          1 3
        /  //                    /
        AGAGTCGACCTGCAGGCATGCAAGCTTGGCACTGGCCGTCG
6260    +---------+---------+---------+    6300
        TCTCAGCTGGACGTCCGTACGTTCGAACCGTGACCGGCAGC
```

autoradiogram of a good probe after the purification. If the reaction is not chased and the dATP concentration used is below the Km of the reaction, a very heterogeneous population of probe molecules is created (Fig. 17). Such a dirty probe would, of course, increase the background and give a low specific signal when used for hybridization. An alternative way would be to synthesize a radioactively labeled RNA probe using the sp6 promoter system.

Vectors such as pSPT 18 and pSPT 19 (Boehringer Mannheim) offer the possibility to clone a defined piece of DNA in a polylinker that is flanked on one side by an SP6 promotor and on the other by a T7 promotor. Such a clone permits a labeled single-stranded RNA probe complementary to either the upper or lower DNA strand to be synthesized simply by using either SP6 or T7RNA polymerase, respectively.

The Choice of the Label

One band in a genomic sequencing track corresponds to approximately 1 femtogram of genomic target DNA, that is a few hundred DNA molecules per band. For this reason the probe should have a high specific activity and the choice of the label should be carefully considered. Biotinylated and other such labeled probes are out of question because of their low power of resolution. Only the radioactive isotopes ^{32}P, ^{35}S and ^{125}I could be used. ^{35}S would be an excellent label because its beta emission has a comparatively low energy, giving less scattering and resulting in very high resolution. Unfortunately it cannot be detected very well with normal laboratory hand monitors which makes it difficult to follow the various steps of the washing of filters after the hybridization procedure. ^{125}I gives a high scattering and therefore could decrease the resolution of the sequence. ^{32}P is probably the best compromise for genomic sequencing. It can be measured easily and the scattering of the beta particles is not too high.

◀

Fig. 16. Map of the linker area of M13mp18

The presence of an M13 "tail" at the 3'-end of the single-stranded probe increases the hybridization signal. After primer extension the double-stranded DNA is cut with the restriction endonuclease PvuII at position 6053. This will generate after strand separation a ca. 180-nucleotide-long tail (tail plus polylinker are indicated by a thick line).

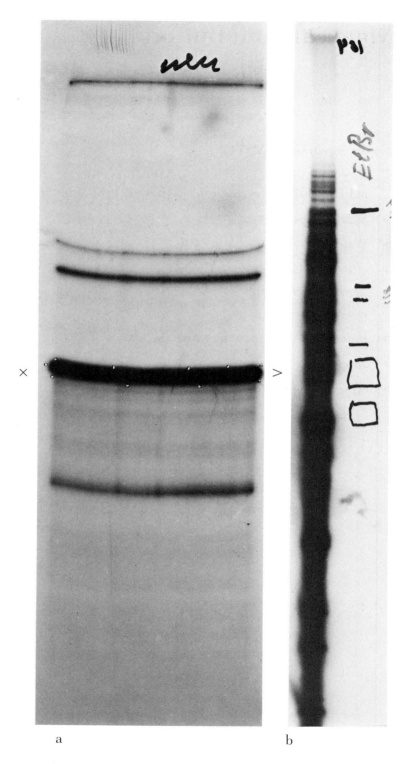

a b

Synthesis of Oligonucleotide Primers

Materials and Buffers

> Single-stranded M13 DNA

> $10 \times$ medium salt buffer: 500 mM sodium chloride; 100 mM Tris-HCl, pH 7.5; 100 mM magnesium chloride; 10 mM dithiothreitol

> Synthetic primer

> Water baths (75, 65, 37, 23° C)

> (^{32}P-alpha) dATP: at least 3000 Ci/mmole; 1 Ci = 37 GBq

> dGTP, dATP, dCTP, dTTP; 10 mM stock solutions in H_2O pH not corrected.

> Polymerase I (Klenow fragment): at least 5 U/μl

> Microfuge

> Appropriate restriction endonuclease

◀

Fig. 17. A comparison of two protocols for the synthesis of a labeled DNA probe

The pulsed chased primed-extension synthesis followed by the digestion with *Pvu*II (a) yields a homogeneous probe (X). The exposure time was only 15 seconds. In (*b*) the concentration of the dATP was below the Km and the synthesis was not chased with unlabeled dATP. This results in a heterogeneous probe. The isolation of the DNA in the correct area of the probe (arrow head) gave only a very small amount of specific labeled probe, which was insufficient for a successful hybridization.

Step-by-Step Procedure

> Add 1 μl (6μg/μl) of M13 single-stranded DNA (ssDNA, see III.11) to 19 μl of water. Add 6 μl of 10 × medium salt buffer according to Maniatis (1982) (500 mM sodium chloride/100 mM Tris-HCl, pH 7.5/100 mM magnesium chloride/ 10 mM dithiothreitol). Add 28 μl (2.5 μg/ml H_2O) of synthetic primer.

> Mix gently by tapping the tube and put the sample at 75° C for 3 minutes. Slowly cool the sample to room temperature by placing the Eppendorf tube in about 80–100 ml of water at 75° C and leaving it to cool down over a period of 30–35 minutes. Place the probe on ice and add 1 μl of 10 mM solutions of dGTP, dCTP, dTTP and 25–28 μl of (^{32}P-alpha) dATP (250 μCi; at least 3000 Ci/mmole). Mix by tapping the tube and spin down a few seconds in microfuge. Add 3 μl (15 U) of DNA polymerase I (Klenow fragment), mix by tapping the tube, give a short spin in a microfuge and incubate the reaction for 7 minutes at 23° C. Add 1 μl of 10 mM dATP, mix by tapping the tube and give a short spin in the microfuge.

> Incubate another 15 minutes at 23° C.

> Stop the reaction by heating the sample for 5 minutes at 65° C. Once it has cooled to room temperature, add 50 units of the appropriate restriction enzyme. Mix by tapping the tube, give a short spin in a microfuge and incubate for 2–3 hours at 37° C.

> The probe is then placed on ice or frozen at this stage if you are not going to continue with its purification immediately (see next chapter).

Notes:

13 Purification of Labeled Single-Stranded Probes

Purification of the probe is essential as nonspecifically labeled molecules increase the background on the hybridized filters. The purification of the restriction-digested primed-extended DNA is performed on a preparative denaturing polyacrylamide gel. The probe can be eluted from the polyacrylamide gel either by diffusion or by electroelution. Of all the different procedures we have tested the worst recovery was obtained by the diffusion method (50% recovery). The recovery by electroelution was usually much better (70–95%). Electroelution was carried out using the Sartorius-collodium bags or in an ISCO-apparatus as described in the following 'Step-by-Step Procedure'. The recoveries range from 70 to 90% in both systems. Nevertheless, the use of bad batches of collodium membranes can result in low recovery of probe. After the electroelution, the eluate has to be filtered (Millex filters) to give a cleaner result in the hybridization. If the probe is not filtered, the background on the hybridized nylon membranes can be very high and may have a speckled appearance on autoradiography. Since the volume of the filtered eluate is too large for the direct hybridization (see 'Step-by-Step Procedure') it must be ethanol precipitated.

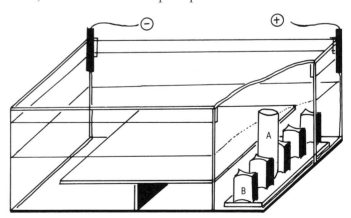

Fig. 18. The electroelution system

A Sartorius collodion bag; *B* holder; inner size of the tank: $20 \times 9 \times 30$ cm; length of the table: 20 cm; height of the table: 4 cm; buffer volume: 3 l. The tank presented here is a standard horizontal electrophoresis apparatus.

Materials and Buffers

> 8 % preparative polyacrylamide sequencing gel

> Sample dye: 94 % formamide; 10 mM Na_2EDTA, pH 7.2; 0.05 % xylenecyanol; 0.05 % bromophenol blue (BPB)

> Water bath (90° C)

> Power supply

> Saran wrap

> X-ray film (any kind)

> Millex-HA 0.45 μm disposable filters (Art. No. SLHA 025BS)

> Syringes (20 ml)

> *E. coli tRNA* (stock solution: 20 μg/μl H_2O)

> 3 M sodium acetate; 5 mM EDTA, pH 5

> Ethanol

> SW 27 rotor (Beckman or equivalent)

> Polyallomer centrifuge tubes (SW 27 rotor)

> Vacuum dryer oven or speed vac

Electroelution using sartorius bags:

> Sartorius collodion bags (15 ml)

> 1 × TBE buffer (10 × TBE: 0.89 M Tris base; 0.89 M boric acid; 0.02 M EDTA; pH 8.3)

> Electroelution chamber (see Fig. 18), a standard horizontal electrophoresis tank

> Corex tubes: 30 ml; silicone-treated

> Hand monitor

Electroelution using the ISCO system (see Fig. 19):

> ISCO apparatus (Model 1750; electrophoretic sample concentrator)

> Dialysis tubings (2.5 cm; Union Carbide)

> Platinum wire

> Power pack (40 mA; 200–300 V)

> 1 × TBE buffer (10 × TBE: 0.89 M Tris base; 0.89 M boric acid; 0.02 M EDTA; pH 8.3)

> Corex tubes (30 ml; silicone-treated)

> *E. coli* carrier DNA (5 μg/μl H$_2$O)

> Bovine serum albumin (crystalline grade)

> Hand monitor

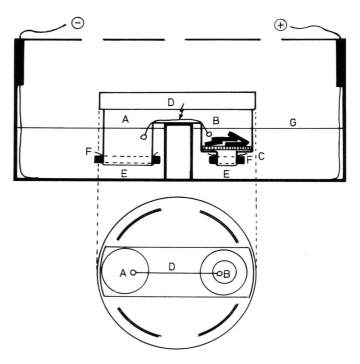

Fig. 19. The ISCO sample concentrator

A Large chamber of the sample concentrator; *B* small chamber of the concentrator; *C* net with gel pieces; *D* platinum wire; *E* dialysis membrane; *F* holder for dialysis membrane; *G* buffer level.

Step-by-Step Procedure

> Prepare an 8% preparative sequencing gel (29:1 acrylamide : bisacrylamide; $400 \times 400 \times 1$ mm; slot size: 60×1 mm; volume of gel solution = 200 ml) according to the protocol in chapter III.4 Preelectrophoresis is carried out at 300 volts for 1 hour.

> Mix the digested labeled DNA with $140 \,\mu l$ of sample dye (94% formamide, 10 mM Na_2EDTA, pH 7.2; 0.05% xylenecyanol; 0.05% bromophenol blue (BPB)).

> Incubate for 1 minute at $90°C$, chill in ice/water and load the probe onto the gel.

> Electrophoresis is carried out at $50-60$ mA constant current until xylenecyanol has migrated about 15 cm from the top of the gel. Remove the top glass plate and cover gel with Saran wrap, avoid trapping air bubbles. Expose the gel for $15-30$ seconds as indicated in Fig. 20.

Note: *If longer exposure time is needed, the probe has a too low specific activity and should be discarded.*

A   Electroelution using Sartorius Bags:

> Rinse a large Sartorius bag (15 ml) for about 30 minutes in 1 liter of distilled water. Excise the piece of gel containing the labeled probe with a scalpel blade (see Fig. 20) and place in the Sartorius bag with 6 ml of distilled water (Fig. 18). Place Sartorius bag in $1 \times$ TBE into an electrophoresis chamber and elute at 200 mA for 10

minutes. Collect the water containing the probe with a Pasteur pipette and place it in a silicone treated Corex tube. Add another 6 ml of water to the contents of the collodion bag and carry on the electroelution as before. Repeat this procedure until as much of the probe is recovered from the gel as possible (recovery: 70–90 %).

B Electroelution using the ISCO sample concentrator:

> Prepare dialysis tubings (2.5 cm large; Union Carbide) as follows: boil tube for 30 minutes in 4 % NaHCO₃, 30 minutes in 5 mM EDTA and finally 10 minutes in distilled water. The dialysis tubing can be autoclaved in 10 mM Tris, pH 8; 1 mM EDTA and kept in the cold.

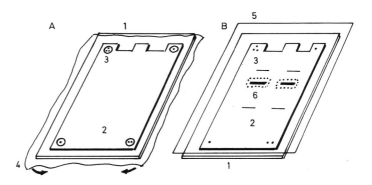

Fig. 20. Exposure of the preparative gel

After the separation of the probe on a preparative gel, the upper glass plate is taken away *(A)*. The gel (2) on the lower glass plate (1), is covered with Saran wrap (4). Several radioactive ink markers (3) are placed on the surface of the Saran wrap. In a dark room a sheet of an X-ray film (5) is placed directly on the gel/ Saran wrap (2) and covered with a second glass plate *(B)*. After 15 to 30 seconds the film is developed. The radioactive markers (3) are visible on the X-ray film and will be used for the precise localization and excision of the probe. Using a needle, several holes are made through the film and gel around the band of interest (6) and the gel is cut out for electroelution.

Purification of Probes

> Cut two pieces of dialysis membrane $(2.5 \times 2.5$ cm; 1.5×1.5 cm; see Fig. 19 E) and fix them as shown in Fig. 19 with the membrane holder F at the elution compartments A and B. Add 5 ml each of distilled water containing $100 \mu g$ carrier $E. coli$ DNA and $50 \mu g$ (1 %) of bovine serum albumin (crystalline grade) to compartment A and B. Connect chamber A and B with a platinum wire (D) and put the elution chamber into the elution tank which is filled to the level G with $1 \times$ TBE buffer. Connect the power supply and run at 40 mA; 200–300 Volts. After ½ hour, the compartments A and B are emptied.

> Put the gel pieces containing the radioactive probe in compartment B (gel pieces can also be placed on a nylon net as indicated in Fig. 19 C). Fill compartments A and B up to level G with distilled water and elute under same conditions as described above. Change polarity for 5 min. Collect the water containing the probe with a Pasteur pipette and transfer it into a silicone-treated Corex tube. Repeat the electroelution as described above until probe is eluted from the gel (recovery 70–90 %).

After electroelution by either systems the probe has to be filtered:

> Filter the labeled probe through a Millex $0.45 \mu m$ membrane which has been prewetted with water (1–2 drops/sec; pressure not too high). Put 10 ml aliquots into polyallomer SW27 Beckman centrifuge tube, add $100 \mu g$ of $E. coli$ carrier tRNA, $\frac{1}{10}$ volume of 3 M sodium acetate, 5 mM EDTA, pH 5 and 2.5 volumes of cold $(-20° C)$ ethanol. Mix by inverting and leave the tubes overnight at $-20° C$. Centrifuge for 1–2 hours at 25 000 rpm. Decant the supernatant, dry the tubes under vacuum and dissolve the sediments in a total volume of $250 \mu l$ $(2 \times 100 \mu l$ and $1 \times 50 \mu l)$ of water. The probe is now ready for use in hybridization.

Notes:

PURIFICATION OF PROBES

IV Trouble-Shooting Guide

Problems

Note: *Many problems occurring in genomic sequencing look the same but may have completely different causes. In this guide we first outline the problems, then the probable causes and finally suggest solutions. The listing will be done according to the chapters (→), as outlined in the flow diagram (p. 22).*

Problem	*High Background in <u>all</u> Sequencing Lanes*

Probable cause
and solution:

→1 **Isolation of genomic DNA:**

– Degradation before, during or after the isolation of the genomic DNA. Use only fresh organs or cells for the DNA isolation, work as quickly as possible and always avoid a shearing of the DNA. In organs, rich in DNAse I, the use of any Mg^{++} in the buffer for the isolation of the nuclei should be avoided; use 1 mM EDTA, 0.1 mM EGTA and replace Mg^{++} with 0.15 mM spermine and 0.5 mM spermidine. The quality of the DNA can be tested by analyzing a few μg of the isolated DNA on an agarose gel just before the restriction digestion: Undegraded DNA should not give a smear on the agarose gel.

→2 **Restriction digest of the genomic DNA:**

– The restriction enzyme may be contaminated with nucleases. This can be tested by a Southern-blot analysis: If the band corresponding to the target fragment is smearing, try to use another batch of restriction enzymes or buy your enzyme from a different supplier.

→3 **Chemical sequencing reactions on restricted DNA:**

– There may be still traces of piperidine in your sample: use a good vacuum for the lyophilizations.

Problem	*More than one Band appearing for each Base in all Sequencing Lanes:*

Probable cause
and solution:

→2 **Restriction digest of genomic DNA:**

– The restriction enzyme does not cut specifically (for example *Eco*RI/*Eco*RI*). On Southern-blot analysis this will result in more bands than expected. Use another restriction enzyme or change the digestion conditions, i.e. salt concentration, enzyme concentration).

— The restriction enzyme is contaminated with other restriction enzymes (endo-, exonucleases) that cut within the target sequence.

— The restriction digest is not complete. This can be tested by Southern-blot analysis: in this case you will find the presence of bands of different lengths and intensities. Make sure that the digestion is complete by using internal controls, such as mixing a few μg of λ DNA with genomic DNA. Incomplete digestion of λ DNA by the restriction enzyme (previously tested for its activity) either means that the buffer is wrong or the DNA preparation is contaminated with an inhibiting substance. DNA prepared from cells treated with dimethysulfate (as in *in vivo* genomic footprinting) may, in certain cases, be resistant to restriction endonucleases. This could be a particular problem if the restriction site in the chromatin is exposed to dimethysulfate and contains guanosine. In such a case it is advisable, (if possible given the limitations imposed by the sequencing strategy) to cleave another restriction site containing no guanosine.

Problem	***More than one specific band per nucleotide***
Probable cause and solution:	**→ 2 Restriction digest of genomic DNA:**

— There is probably more than one gene or pseudogene per haploid genome with a similar restriction fragment. Try to use another restriction endonuclease. If this does not help, this problem may be impossible to solve.

Problem	*Random ladders in one or more of the sequencing lanes*

Probable cause and solution:	→ 3	**Chemical sequencing reactions on restricted DNA:**

— Presence of too many cuts within the DNA due to an excessive chemical modification: For example concentration of DNA-modifying (Maxam & Gilbert) chemicals was too high, temperature too high, the reaction time too long, the concentration of genomic DNA too low usually as a result of an incorrect OD measurement caused by other UV-absorbing material in the preparation or the DNA was only partially solubilized. Control the volumes of the modifying chemicals very carefully using micropipettes. Carefully control the temperature and reaction time. If this still fails increase the amount of genomic DNA per sequencing reaction.

Problem	*Visible T's in the (C)-lane*

Probable cause and solution:	→ 3	**Chemical sequencing reactions on restricted DNA:**

— Hydrazine has become oxidized; replace it.

Problem	*Visible C's in the (T)-lane*

Probable cause and solution:	→ 3	**Chemical sequencing reactions on restricted DNA:**

— Nonspecific modifications in the pyrimidines: increase DNA concentration for the potassium permanganate reaction.

Problem

Weak T's in the (C & T)-lane

Probable cause
and solution:

→ 3 **Chemical sequencing reactions
on DNA:**

– The presence of residual salt after ethanol precipitation has inhibited the T-reaction. Redissolve dried DNA in 0.3 M sodium acetate/0.5 mM sodium-EDTA, pH 5 and precipitate DNA once more with 3 volumes of ethanol. Remove all of the supernatant using a drawn-out capillary.

Problem

*Extra bands corresponding to G within
the (C)-and/or (T & C)-lane*

Probable cause
and solution:

→ 3 **Chemical sequencing reactions
on restricted DNA:**

– Piperidine cleaves at guanines after a reaction between guanine and hydrazine at low pH; therefore hydrazine-stop solution should be ice-cold (0° C) before use. Chill the sample as quickly as possible after the hydrazinolysis by adding pre-chilled stopping buffer and ethanol. Use cooled centrifuge.

Problem

Poor resolution of bands, bands distorted

Probable cause
and solution:

→ 4 **Separation of reaction product on
a sequencing gel:**

– Samples are diffusing during loading the gel; change the ratio of sample dye to aqueous probe.

– The concentration of urea in the wells is too high; clean the wells very carefully with a stream of buffer (use a syringe) before loading the probe.

– The bottoms of the wells are uneven or damaged; after the gel has polymerized the comb should be removed very carefully, and when cleaning the wells, direct contact with the gel should be avoided.

- There are still gel particles or other 'dirt' in the wells; as described above, the wells should be cleaned thoroughly before loading the probes.

- Diffusion of the bands during electrophoresis; run the gel at a higher temperature by increasing the current or run the gel in a heated mantle.

- The DNA is not completely in solution. Dissolve the DNA very carefully in water before adding the sample dye; heat the sample for 1 to 2 minutes at $90-95°C$ before quick chilling and loading onto the wells.

- Air bubbles trapped at the bottom of the gel (anode chamber); This may result in heterogeneous electrical field and a distortion of the sequencing lanes. Remove air bubbles with a stream of buffer and make sure that lower end of the gel is not directly above the anode-platinum wire (because of the production of O_2-bubbles).

- Structural intra- or intermolecular effects of DNA molecules during electrophoresis. Denaturation is incomplete. Increase current and/or urea concentration (from $7\,M$ to $8\,M$ urea).

- The buffer capacity is too low; use only $1 \times$ TBE and change buffer at least every 6 hours or more often, or increase the volume of the electrophoresis chambers.

Problem

No differences in the electrophoretic mobilities between the (G)- or (G + A)- and the following (C)- or (C + T)-bands

Probable cause and solution:

→ 4 **Separation of reaction products on a sequencing gel:**

- The buffer capacity is too low. Electrolysis of the buffer results in a pH change (anode: acid; cathode: basic); use a large volume of $1 \times$ TBE and change the buffer as described in the 'Step-by-

Step Procedure'. The pH can be easily tested by putting 'non-bleeding pH-strips' directly into the buffer tanks. In any case, avoid recycling of the buffer as it will ultimately result in an increase in the background.

Problem	***Single bands are distorted or lost***
Probable cause and solution:	→5 **Electrotransfer to nylon membranes:**

– Gas bubbles were trapped in the transfer system; this results in an uneven electrical resistance; avoid Scotch Brite like pads; use only thick, pre-soaked filter paper; use a heavier weight on top of the transfer system.

Problem	***Weaker signals at the top of the filter than at the bottom***
Probable cause and solution:	→5 **Electrotransfer to nylon membranes:**

– The transfer of larger DNA fragments is incomplete. Increase the time of electrotransfer.

Problem	***No signals on autoradiography***
Probable cause and solution:	→5 **Electrotransfer to nylon membranes:**

– The polarity in the transfer system was wrong. Also check carefully the order of assembly of the gel, filter, anode and cathode (see Fig. 8). Make sure that the layer of filter paper on top of the nylon membrane is not in direct contact with the lower layer of filter paper.

Problem	*Areas without any signal* (see Fig. 21/3)
Probable cause Sand solution:	→5 **Electrotransfer to nylon membranes:** – Air bubbles are trapped below gel and membrane. Place the gel on Whatman sheet only when there is no visible buffer on the surface anymore, as described in the 'Step-by-Step Procedure'. Eliminate any air bubbles before putting the membrane onto the gel.
Problem	***Double bands (one weak and the other one strong) within the sequencing ladder***
Probable cause Sand solution:	→5 **Electrotransfer to nylon membranes:** – The membrane had made contact more than once with the gel. Once the membrane is touching the gel, do not re-position it.
Problem	*Overall weak signals*
Probable cause and solution:	→4 **Separation of reaction products on a sequencing gel:** – There was not enough DNA loaded onto the gel. Check carefully that all the DNA of the sample is dissolved. If the suggested volume is too small, increase by adding another few microliters of sample mixture (1 part water, 2 parts sample buffer). Ensure that all DNA in the Eppendorf tube is loaded into the well. Because of the complexity of the genome to be studied, you need more DNA than indicated in the 'Step-by-Step Procedure'. In extreme cases you may need to enrich the gene by preparative agarose gel electrophoresis.
	→6 **Immobilization of DNA on a nylon membrane:** – Wrong UV-dose resulting in too high binding of genomic DNA onto the membrane. Check each

new batch of filters for their capacity to bind DNA upon UV-irradiation. Do not forget that the binding should be approx. 30–50% of the input DNA (see Fig. 12, chapter III.6). Should you change the UV lamps while still using the same batch of membrane, it will be necessary to repeat membrane calibration (p. 70).

→ 7 Prehybridization and hybridization of immobilized DNA with labeled single-stranded DNA probes:

– The hybridization conditions are too stringent. Decrease the hybridization temperature. If there is still no improvement, use a probe with higher specific radioactivity and/or decrease the volume of the hybridization chamber by using a thinner nylon fishing line.

– The walls of the hybridization chamber have not been treated with silicone and the radioactive probe is attached to the glass.

– The composition of the hybridization buffer is wrong. Check the salt concentration and pH.

– Have a look at the base composition of the hybridization probe and choose the labeled nucleotide accordingly. If necessary use two different nucleotides. After primed-extension cut the probe with a restriction enzyme that cleaves in the 5'-flanking area of the insert. The M13 tail should not exceed 200 nucleotides.

→ 8 Processing of the hybridized filters:

– The washing conditions of the filters are too stringent. Filter bound radioactivity decreases very quickly from one wash to the next; therefore decrease the temperature of the buffer and/or increase the salt concentration. Use only a small volume of the washing buffer, as the temperature of preincubated buffer should decrease rapidly to room temperature.

– The hybridization probe is too short. In our experience the best results were seen using a probe at least 90–120 nucleotides long (this size is without the M13 tail). However, it should be noted that too long a probe can also affect the resolution of the sequence that can be read.

Problem

Weak signals in one or more sequencing lanes

Probable cause and solution:

→4 **Separation of reaction product on a sequencing gel:**

– There was not enough DNA loaded onto the gel. Make sure that the sample has been completely dissolved, and load the sample very carefully avoiding any turbulence.

Problem

The strength of the hybridization signal varies across the filter

Probable cause and solution:

→6 **Immobilization of DNA on a nylon membrane:**

– Surface of the membrane has been partially altered: only touch the membrane with clean gloves. In the worst of cases the binding capacity for DNA will change within the batch of filter; forget that batch and try a new one.

Problem

More or less strong interbands in all sequencing lanes (see Fig. 21/2)

Probable cause and solution:

→3 **Chemical sequencing reaction on restricted DNA:**

– The level of chemical modifications of the bases in the DNA was too high, resulting in too many cuts per unit length of DNA. If the concentration of DNA suggested in the 'Step-by-Step Procedure'

(p. 45) was not sufficient, increase the amount of DNA per reaction to 75 μg. Decreasing the temperature and/or concentration of DNA-modifying chemicals of the sequencing reactions is not recommended.

→ 7 Prehybridization and hybridization of immobilized DNA with labeled single-stranded DNA probes:

– The hybridization probe is too long; probe is hybridizing with genomic fragment with an unequal start (due to double cuts within the probe area of the genomic DNA).

– The sequence of the probe molecule is not unique and cross-hybridizes with other DNA fragments. Test the sequence of the probe by means of a computer-assisted sequence comparison and by Southern-blot analysis.

– The stringency of hybridization is too low. Increase hybridization temperature.

– An old probe has not been completely removed before rehybridizing the filter.

→ 8 Processing of the hybridized filters:

– There is a cross-hybridization with other sequences; the washing conditions are not stringent enough; increase the temperature or decrease the salt concentration in buffer 2 (p. 87).

– Presence of labeled hybrid molecules on the filter before a second hybridization with the probe of the second strand. Increase the temperature of the washing buffer or use pure water to get rid of the remaining probe.

→ 10 Cloning of DNA probe into M13:

– Phage contamination (cloning of symmetrical DNA fragments into the same vector). Take only well-separated positive phage plaques from plates with a relatively low plaque density.

Problem	*Hot spots dispersed over the whole filter membrane* (see Fig. 21/1)

Probable cause and solution:

→7 **Prehybridization and hybridization of immobilized DNA with labeled single-stranded DNA probes:**

— The probe is contaminated with particles absorbing radioactive molecules. It is important that the probe is filtered as described in the 'Step-by-Step Procedure' before use and that the glass plates of the hybridization chamber are spotlessly clean and treated with silicone.

— The various substances in the prehybridization and hybridization buffer are not completely in solution; dissolve the chemicals of the hybridization buffer completely before adding the denatured radioactive probe.

Problem	*High background on filter membrane* (see Fig. 21/4)

Probable cause and solution:

→7 **Prehybridization and hybridization of immobilized DNA with labeled single-stranded DNA probes:**

— The transfer and/or immobilization of DNA onto the membrane was not optimal, due to a bad batch of membranes. Take another batch of membranes and ask for your money back!

— The composition of the hybridization buffer was wrong. Make up the buffer exactly according to the protocol. Be careful when using other hybridization buffers as they often show a decreased upper limit for the concentration of the labeled probe. Exceeding that limit will result in a high background which is no longer possible to get rid of.

— When two or more filters are hybridized in the type of chamber described here. Hybridize only one filter per chamber (Fig. 21/7).

- The membrane was not soaked with water and/or prehybridized before adding the radioactive hybridization mixture.

→ 8 Processing of the hybridized filters:

- Wrong washing conditions. Make an additional wash of the membrane with buffer 2 (see buffers, p. 87). If the background does not decrease, use a more stringent buffer condition.

- Surface of the membrane has been altered. Repeat the experiment with a new filter.

- The filter has been contaminated with radioactivity-absorbing material. Store membranes only between filter papers with a smooth surface.

→ 9 Autoradiography and photography:

- Film has been overflashed. X-ray film absorption (540 nm) upon preflashing should range between 0.1 and 0.2 OD; if the intensity of the flash is too high the background will increase drastically.

→ 13 Purification of labeled single-stranded probes:

- Electroelution of the probe with Sartorius bags was too long. After long elutions, an as yet unknown substance, is also eluted, which is not subsequently removed by filtration. This substance influences the background on the filter membranes.

Problem	***X-ray film shows big silver grains***
Probable cause and solution:	**→ 9 Autoradiography and photography:**

- The film is too sensitive; use another film with higher silver-halide concentration (example: Kodak, X-OMAT S).

Problem	*Smears on X-ray film*
Probable cause and solution:	**→9 Autoradiography and photography:**

- Saran wrap cover is contaminated. Use only clean gloves when covering the membrane with Saran wrap.

- Water drops between Saran wrap and film and/or between film and intensifying screen; either the film should be developed immediately after removal from the freezer, or otherwise it has to be completely dried before it is developed.

- Electrostatic discharges can under certain circumstances give lightning-like signals on the film (see Fig. 21/9). To avoid this take the film off the membrane slowly. Avoid any direct contact of the film with sticky material, such as Scotch tape.

Problem	*Low specific activity of the hybridization probe (labeled ssDNA probe)*
Probable cause and solution:	**→12 Synthesis of oligonucleotide primers and single-stranded labeled probes:**

- Inactive Klenow enzyme. Store enzymes at $-20°$ C and keep in ice before use.

- Primer concentration is not high enough. For $6\,\mu g$ of single-stranded DNA use at least $70\,ng$ of oligo primer (for an insert between 100 and 300 nucleotides).

- Poor annealing reaction. For annealing DNA, the sample should first be heated up to about $75°$ C, and then slowly allowed to cool down to room temperature before being placed on ice (further manipulation should follow as soon as possible).

- The restriction digest did not work well. If different enzymes other than those described are used for the cleavage of the probe, be sure that they work in medium salt buffer. Otherwise adapt the salt concentration accordingly.

- Exonuclease activity. Incubate sample after primed-extension for at least 5 min at 65° C to destroy polymerase.

- Impure deoxynucleotide triphosphate used for the primed-extension. Replace the deoxyribonucleotide triphosphates; store them in aliquots at −70° C. Partial degradation of the labeled dATP. Use a fresh batch of labeled nucleotides.

- The DNA is not clean enough, and contains substances that inhibit the polymerase. Purify the DNA either by ethanol precipitation or by reverse-phase chromatography.

Problem

Loss of the probe during purification

Probable cause and solution:

→ **13 Purification of labeled single-stranded probes:**

- Parts of the probe have been lost during electroelution. Use only Sartorius bags that have been pre-incubated in distilled water for at least 0.5 hour. It is also important to change the eluate every 6–10 minutes because once the probe sticks in the wall of the Sartorius bag it cannot be recovered). Occasionally we have also found bad batches of Sartorius bags; replace the whole batch.

- Parts of the probe are lost during the ethanol precipitation. Sodium acetate and *E. coli* tRNA (carrier) should be added to the eluate to a final concentration of 0.3 M and 10 μg/ml respectively before precipitating with 2.5–3 volumes of ethanol. The precipitated ssDNA probe is then collected by ultracentrifugation (p. 123).

Figs. 21
Examples to the trouble-shooting guide

The following pictures show some possible errors occurring during genomic sequencing:

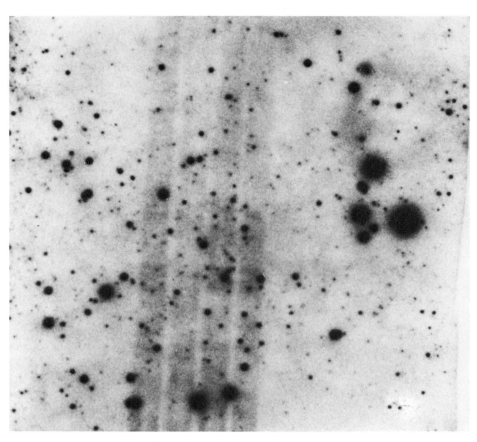

21/1

Hybridization probe was not filtered. This may result in hot spots dispersed over the whole filter membrane (compare p. 137). In addition, the DNA was partially degraded due to a bad batch of restriction-enzyme (compare with p. 127).

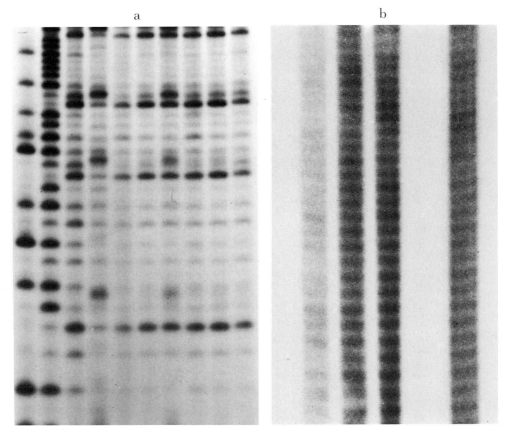

a   b

21/2

Examples *a* and *b* show interbands in all sequencing lanes. In both experiments the wrong sequencing-reaction conditions were used: the quantity of genomic DNA for the sequencing reactions was too low. This resulted in a too high chemical modification of the bases. The sequence of *a* can still be read whereas *b* shows a clear random ladder with a high background between the bands (compare p. 41 and 129).

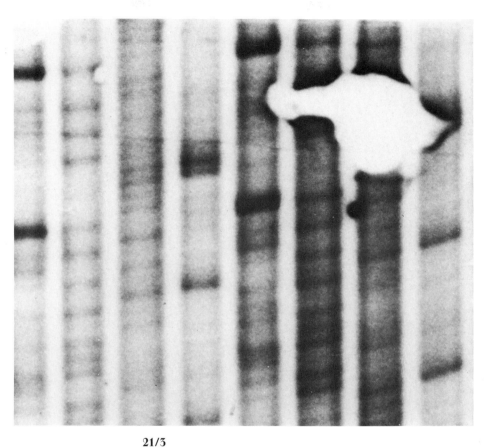

21/3

The area without any signal was due to an air bubble trapped below the gel during the electrophoretic transfer (see p. 133).

21/4

High background due to a hybridization buffer containing
formamide with a too high concentration of labeled DNA probe.

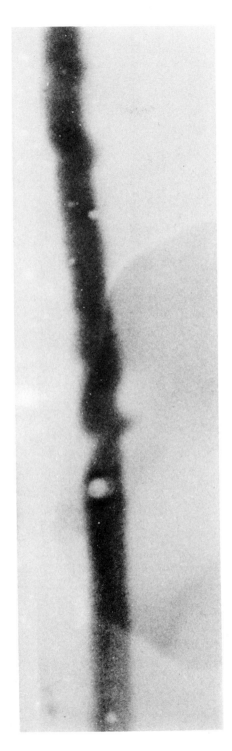

21/5

The use of an inadequate electrotransfer system may result in distorted sequencing lanes with no resolution of the bands. For this example we used a commercially available vertical system with the Scotch brites pads. The contact between the different elements of the assembly system was not tight enough and the distance between the electrodes was too great.

X

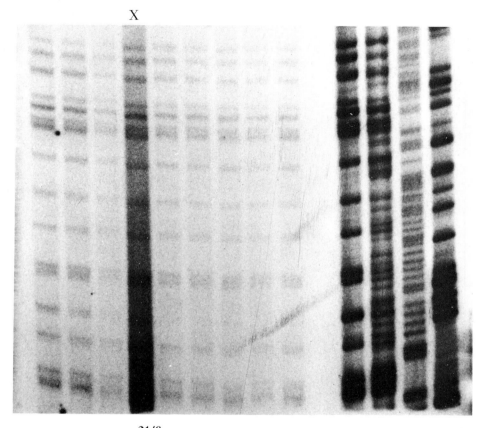

21/6

The DNA of the indicated sequencing lane (X) was partially degraded.

a   b

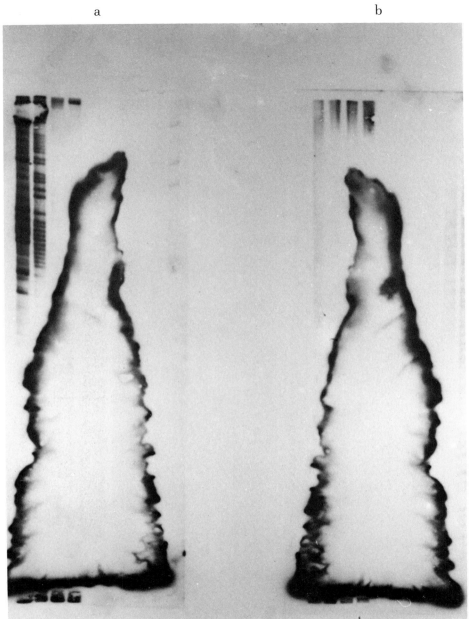

21/7

The two filters *a* and *b* were hybridized in the same hybridization chamber (0.35 mm ∅ fishing line). The areas of contact between the two filters show an increase of the marginal background with no hybridization in the center.

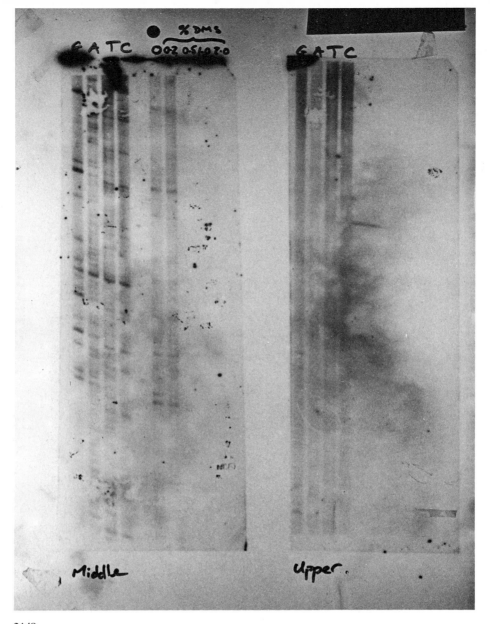

21/8

This example shows several mistakes in one experiment.

1) The filter was not covered completely with hybridization buffer within the hybridization chamber. This results in a very strong and irreversible signal at the top of the filter.

2) The filter was not washed enough, this results in a high general background.

3) The black spots are due to the Saran wrap which was not completely dry before developing the film.

4) The areas within the sequencing lanes without any signal are due to air bubbles trapped below the gel during the electric transfer.

21/9

Star-like background on the X-ray
film due to electrostatic discharges.

V Appendix

A Suppliers of Special Items and Construction of Equipment

This part is also organized according to the flow diagram shown on p. 22.

→4 Separation of reaction product on a sequencing gel:

– Glass plates (1000 × 330 × 5 mm) for genomic sequencing gels: There is no special glass needed for genomic sequencing and suitable plates are available in any glaziery.

– Spacer (PVC): 2× (1 × 20 × 1000 mm) and 1× (1 × 20 × 40 mm); Comb (PVC): 15 teeth (1 × 5 × 25 mm; space between the teeth 3 mm).
Supplier: High precision spacers and combs are available from Zabona AG (Labor- und Medizinal-Technik), Mattenstrasse 16, 4058 Basel, Switzerland.

– High voltage power supply: based on experience we highly recommend the LKB 2297 power supply (Macrodrive 5):

Regulation	Constant voltage, constant current and constant power.
Output	Voltage: 0–5000 V, current: 0–150 mA, power: 0–200 W

Mains voltages 100–132 V, 198–264 V; 50 or 60 Hz.
Supplier: LKB-Produkter AB, Box 305, S-161 26 Bromma, Sweden.

– Stands for long gels (as described in the text).
Supplier: Texas Radiological Instrument Co., P. O. Box 421, Friendswood, Texas 77 546, USA.

→ 5 Electrotransfer to nylon membranes:

– Power packs for high current: at least 2 A. There are only a few different types available on the market. We have had good results with the Biotec NG 60/5 power pack:
Current: up to 5 A, voltage: 5–60 V
Mains voltages: 110 V, 220 V; 50, 60 Hz or to need of local supply
Price: about 1080 Swiss francs or 800 US $
Supplier: Biotec AG, Sissacherstrasse 24, 4460 Gelterkinden, Switzerland.

– An alternative power pack is the BioRad-Model 250/2.5
Current: up to 2.5 A, voltage: up to 250 V
Price: 1500 Swiss francs
Supplier: BioRad, 1414 Harbour Way South, Richmond, California 94 804, USA.

– Steel plate sieve electrodes (1.5 × 150 × 400 mm; V4A (BN2); see Fig. 22). Such steel plates should be available in bigger hardware stores and can easily be converted to electrodes by attaching a cable (4–5 mm^2) as shown in figure.

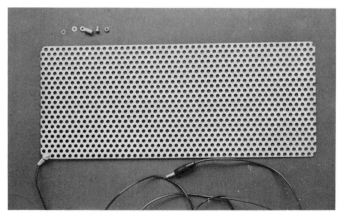

Fig. 22. Steel plate sieve electrode (1.5 × 150 × 400 mm)

The cable is fixed by using a connecting tube, two washers and an assembling screw, as indicated in the figure.

- Steel plate sieve electrodes ready for use (size: $1.5 \times 150 \times 400$ mm)

  Supplier: Biotec AG, Sissacherstrasse 24, 4460 Gelterkinden, Switzerland, Tel. 0619944 66. Price: about 45 Swiss francs or approx. 30 US $.

- Steel plate sieves (german translation: gelochtes Stahlblech; sold as raw material)
  Suppliers: Gysin & Cie, Fabrik für gelochte Bleche, 4622 Egerkingen, Switzerland, or Small Parts, Inc., P. O. Box 381736, Miami, Florida 33 238–1736, USA.

→ 6 Immobilization of DNA on a nylon membrane:

- UV-chamber: The UV-chamber used in our laboratory is homemade. Plans are given in Fig. 23 & 24. The chamber contains two UV tubes (TUV 15W, G 15 T 8; germicidal lamps) from Philips, Holland.

→ 12 Synthesis of Oligonucleotide primers and single-stranded probes:

- 3'-Deoxyadenosine 5'-(alpha-^{32}P)triphosphate (sodium salt, 10 Ci/mmol; 1 Ci = 37 GBq)
  Supplier: Amersham International plc, White Lion Road, Amersham, Buckinghamshire HP7 9LL, UK.

- Sartorius Collodion-Bags (SM 132 00 E; length: 57 mm, diameter: 15 mm).
  Suppliers: Sartorius GmbH, Postfach 19, Weender Landstraße 94–108, D-3400 Göttingen, FRG, or Sartorius Filters Inc, 26 575 Corporate Avenue, Hayward, California 94 545, USA.

→ 13 Purification of labeled single-stranded probes:
- ISCO apparatus: Model 1750; Electrophoretic sample concentrator.
  Supplier: ISCO, Inc. P. O. Box 5347, 4700 Superior Street, Lincoln, Nebraska 68 505, USA (Telex 48-4353).

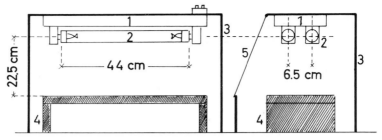

Fig. 23. The UV-chamber

1) lamp holder; 2) UV lamp; 3) metal box; 4) table; 5) Perspex (plexi) glass.

Fig. 24. General view of the UV chamber

It is an old inoculation box converted into an UV chamber. The front side of the UV chamber can be opened completely *(a)*, or closed by means of a plexiglass window *(b)*.

B   Determination
of DNA Concentration

1 DNA Determination by the
diphenylamine reaction (Burton, 1955):

- Dilute the nucleic acid extract with 0.5 M $HCLO_4$ to a final concentration 0.02–0.25 μmol of DNA-P per ml.

- Prepare diphenylamine reagent: dissolve 1.5 g of steamdistilled diphenylamine in 100 ml of redistilled acetic acid. Add 1.5 ml of H_2SO_4 and mix well. Store solution in the dark. Before use, add 0.10 ml of aqueous acetaldehyde (16 mg/ml) for each 20 ml.

- 1–2 ml is mixed with 2 volumes of diphenylamine reagent.

- Prepare a blank, containing 0.5 M $HCLO_4$ but no DNA.

- Incubate samples at 30° C for 16–20 hours in the dark.

- Measure the optical density at 600 nm against the blank and compare with samples containing known amounts of DNA. Alternatively, a more precise standard curve can be prepared using serial dilutions of deoxyribose.

2 Fluorescence assay for DNA determination
(Fiszer-Szafarz et al., 1981):

- Dissolve 0.3 g DABA (3.5.-diaminobenzoic acid) in 1 ml of 4 M *HCl* by vortexing.

- Add 150 mg of charcoal and vortex several times for 15–30 seconds during a 5–10 minute period.   ,4 - .5 filter

- Centrifuge for 1 minute in a benchtop centrifuge.

- Filter supernatant through a polycarbonate filter. (Keep this solution in the dark and use within 2 hours.)

- Add DNA (in water) to DABA solution:
1 μl (10–100 ng) of DNA to 10 μl of DABA solution or 20 μl of DNA (100–500 ng) to 20 μl of DABA solution.

- Mix and incubate for 35 minutes at 65° C.

- Put the samples on ice for 5 minutes.

- Add to each sample 1 ml of 1 M HCl and mix gently by inverting the tubes.

- Centrifuge in a benchtop centrifuge for 5 minutes and transfer the supernatant to a quartz cuvette. Determine the fluorescence in a fluorimeter using excitation and emission wavelengths of 408 and 508 nm, respectively.

- Calculate the quantity of DNA using a standard curve.

Flow Diagram

Total Genomic DNA   The Sequence   Probe for Indirect
Standard   End-Labeling
(Cloned DNA)   (Single-Stranded
DNA or RNA Probes)

___

1. Isolation of
   Genomic DNA
   ▼
2. Restriction Digest   Restriction Digest   10. Cloning of DNA
   of Genomic DNA   of Cloned DNA   Probe in M13

___

                                   ▼

3. Chemical Sequencing Reactions   11. Large-Scale
   on Restricted DNA   Preparation of
       ▼   Cloned DNA in M13
4. Separation of Reaction Products   ▼
   on a Sequencing Gel   12. Synthesis of Oligo-
       ▼   nucleotide Primers
5. Electrotransfer to Nylon Membranes   and Labeled Single-
       ▼   Stranded Probes
6. Immobilization of DNA on a Nylon   ▼
   Membrane   13. Purification of
       ▼   Labeled Single-
7. Prehybridization and Hybridization   Stranded Probes
   of Immobilized DNA with Labeled
   Single-Stranded DNA Probes   ◄─────┘
       ▼
8. Processing of the Hybridized Filters
       ▼
9. Autoradiography and Photography

VI Bibliography

Anderson, M. L. M., and Young, B. D., Quantitative filter hybridization. Nucleic acid hybridization. Eds B. D. Hames and S. J. Higgins. IRL PRESS, Oxford/Washington DC 1985.

Anderson, N. L., Nance, S. L., Pearson, T. W., and Anderson, N. G., Specific anti-serum staining of two-dimensional electrophoretic patterns of human plasma proteins immobilized on nitrocellulose. Electrophoresis 3 (1982) 135–142.

Becker, P. B., Gloss, B., Schmid, W., Straehle, U., and Schuetz, G., In vivo protein-DNA interactions in a glucocorticoid response element require the presence of the hormone. Nature 324 (1986) 686–688.

Bonner, T. I., Brenner, D. J., Neufield, B. R., and Britten, R. J., Reduction in the rate of DNA reassociation by sequence divergence. J. molec. Biol. 81 (1973) 123–135.

Burton, K., A study of the conditions and mechanism of the diphenylamine reaction for the colorimetric estimation of deoxyribonucleic acid. Biochem. J. 62 (1956) 315–323.

Cannon, G., Heinhorst, S., and Weissbach, A., Quantitative molecular hybridization on nylon membranes. Analyt. Biochem. 149 (1985) 229–237.

Church, G. M., and Gilbert, W., Genomic Sequencing. Proc. natl Acad. Sci. USA 81 (1984) 1991–1995.

Church, G. M., Ephrussi, A., Gilbert, W., and Tonegawa, S., Celltype-specific contacts to an immunoglobulin enhancer in nuclei. Nature 313 (1985) 798–801.

Craig, L., Techniques for the study of peptides and proteins by dialysis and diffusion, in: Methods in Enzymology XI, pp. 870–905. Ed. C. H. W. Hirs. Academic Press, New York/London 1967.

Darnell, J., Lodish, M., and Baltimore, D., Molecular Cell Biology. Scientific American Book (1987).

Ephrussi, A., Church, G. M., Tonegawa, S., and Gilbert, W., B lineage-specific interactions of an immunoglobulin enhancer with cellular factors in vivo. Science 227 (1985) 134–140.

Fiszer-Szafarz, B., Szafarz, D., and Guevare de Murillo, A., A general, fast, and sensitive micromethod for DNA determination; application to rat and mouse liver, rat hepatoma, and yeast cells. Analyt. Biochem. 110 (1981) 165–170.

Friedmann, T., and Brown, D. M., Base-specific reactions useful for DNA sequencing methylene blue-sensitized photooxidation of guanine and osmium tetraoxide modification of thymine. Nucl. Acids Res. 5 (1978) 615–622.

Gibson, W., Protease-facilitated transfer of high-molecular-weigth proteins during electrotransfer to nitrocellulose. Analyt. Biochem. 118 (1981) 1–3.

Giniger, E., Varnum, S. M., and Ptashne, M., Specific DNA binding of GAL4, a positive regulatory protein of yeast. Cell 40 (1985) 767–771.

Howley, P. M., Israel, M. F., Law, M.-F., and Martin, M. A., A rapid method for detecting and mapping homology between heterologous DNA evaluation of polyoma virus genomes. J. biol. Chem. 254 (1979) 4876–4883.

Jelley, J. V., Cerenkov radiation and its application. Pergamon Press, London 1958.

Laskey, R. A., and Mills, A. D., Enhanced autoradiographic detection of ^{32}P and ^{125}I using intensifying screens and hypersensitized film. FEBS Lett. 82 (1977) 314–316

Laskey, R. A., Radioisotope detection by fluorography and intensifying screens. Review 23 (1984) Amersham International.

Lohr, D., The salt dependence of chicken and yeast chromatin structure. J. biol. Chem. 261 (1986) 9904–9914.

Marmur, J. G., and Doty, P., Thermal renaturation of deoxyribonucleic acids. J. molec. Biol. 3 (1961) 584–594.

Martin, K., Huo, L., and Schleif, R. F., The DNA loop model for *ara* repression: *AraC* protein occupies the proposed loop sites in vivo and repression-negative mutations lie in these same sites. Proc. natl Acad. Sci. USA 83 (1986) 3654–3658.

Maxam, A. M., and Gilbert, W., Sequencing end-labeled DNA with base-specific chemical cleavages. Meth. Enzymol. 65 (1980) 499–560.

McCellan, T., and Ramshaw, J. A. M., Serial electrophoretic transfers, a technique for the identification of numerous enzymes from single poly-acrylamide gels. Biochem. Genet. 19 (1981) 647–654.

Meinkoth, J., and Wahl, G., Hybridization of nucleic acids on solid support. Analyt. Biochem. 138 (1984) 267–284.

Mendelsohn, S. L., and Young, D. A., Efficacy of sodium dodecyl sulfate, diethyl pyrocarbonate, proteinase K and heparin using a sensitive ribonuclease assay. Biochim. biophys. Acta 519 (1978) 461–473.

Messing, J., New M13 vectors for cloning. Meth. Enzymol. 101 (1983) 20–78.

Nick, H., and Gilbert, W., Detection in vivo of protein-DNA interactions within the lac operon of Escherichia coli. Nature 313 (1985) 795–797.

Nick, H., Bowen, B., Terl, R. J., and Gilbert, W., Detection of cytosine methylation in the maize alcohol dehydrogenase gene by genomic sequencing. Nature 319 (1986) 243–246.

Phillips, C. A., Smith, A. G., and Hahn, E. J., Recent developments in autoradiography, in: Proceedings from the Second International Symposium on the Synthesis and Application of Isotopically Labeled Compounds, pp. 189–194. Ed. R. R. Muccino. Elsevier, New York 1986.

Rubin, C. M., and Schmid, C. W., Pyrimidine-specific chemical reactions useful for DNA sequencing. Nucl. Acids Res. 8 (1980) 4613–4619.

Saito, I., Sugiyama, H., Matsuura, T., Ueda, K., and Komano, T., A new procedure for determining thymine residues in DNA sequencing. Photoinduced cleavage of DNA fragments in the presence of spermine. Nucl. Acids Res. 12 (1984) 2879–2885.

Saluz, H. P., and Jost, J. P., Optimized genomic sequencing as a tool for the study of cytosine methylation in the regulatory region of the chicken vitellogenin II gene. Gene 42 (1986) 151–157.

Saluz, H. P., Jiricny, J., and Jost, J. P., Genomic sequencing reveals a positive correlation between the kinetics of strand-specific DNA demethylation of the overlapping estradiol/glucocorticoid-receptor binding sites and the rate of avian vitellogenin mRNA synthesis. Proc. natl Acad. Sci. USA 83 (1986) 7167–7171.

Shuttleworth, A. D., An improved electrophoretic transfer (electroblotting) apparatus. Electrophoresis 3 (1984) 178–179.

Singh, L., and Jones, K. W., The use of heparin as a simple cost-effective means of controlling background in nucleic acid hybridization procedures. Nucl. Acids Res. 14 (1984) 5627–5638.

Smith, A. G., Phillips, C. A., and Hahn, E. J., X-ray films: suppression of reciprocity failure by astronomical techniques. J. Imag. Technol. 11 (1985) 27–32.

Smith, A. G., Phillips, C. A., Hahn, E. J., and Leacock, R. J., Hypersensitization and astronomical tests of X-ray films: AAS Photo-Bulletin. 39 (1985) 8–14.

Southern, E., Detection of specific sequences among DNA fragments separated by gel electrophoresis. J. molec. Biol. 98 (1975) 503–517.

Sproat, B. S., and Gait, M. J., Solid-phase synthesis of oligodeoxyribonucleotides by the phosphodiester method. Oligonucleotide synthesis. Ed. M. J. Gait. IRL press, Oxford, Washington DC 1984.

Towbin, J., Staehelin, T., and Gordon, J., Electrophoretic transfer of proteins from acrylamide gels to nitrocellulose sheets: procedure and some applications. Proc. natl Acad. Sci USA 76 (1979) 4350–4354.

Tsai, S. Y., Roop, D. R., Stumpf, W. E., Tsai, M. J., and O'Malley, B. W., Evidence that deoxyribonucleic acid sequences flanking the ovalbumin gene are not transcribed. Biochemistry 19 (1980) 1755–1760.

Vaessen, R. T. M. J., Kreike, J., and Groot, G. S. P., Protein transfer to nitrocellulose filters. FEBS Lett. 124 (1981) 193–196.

Walker, P. R., and Sikorska, M., Modulation of the sensitivity of chromatin to exogeneous nucleases: Application for the apparent increased sensitivity of transcriptionally active genes. Biochemistry 25 (1986) 3839–3845.

Zinn, K., and Maniatis, T., Detection of factors that interact with the human beta-interferon regulatory region in vivo by DNAse I Footprinting. Cell 45 (1986) 611–618.

VII Index